Words and Their Meaning

Howard Jackson

 Routledge
Taylor & Francis Group

LONDON AND NEW YORK

First published 1988 by Pearson Education Limited

This edition published 2013 by Routledge

2 Park Square, Milton Park, Abingdon, Oxon OX14 4RN
711 Third Avenue, New York, NY 10017, USA

First issued in hardback 2017

*Routledge is an imprint of the Taylor & Francis Group,
an informa business*

British Library Cataloguing in Publication Data
Jackson, Howard
    Words and their meaning. – (Learning
    about language).
    1. Semantics
    I. Title   II. Series
    412   P325
ISBN 13: 978-1-138-41391-7 (hbk)
ISBN 13: 978-0-582-29154-6 (pbk)

Library of Congress Cataloging in Publication Data
Jackson, Howard, 1945–
    Words and their meaning.

    (Learning about.language)
    Bibliography: p.
    Includes index
    1. English language — Semantics.   2. English
language — Lexicography.   I. Title   II. Series.
PE1585.J33   1988   412   87–3070

Set in Linotron 202 11/12pt Bembo

Contents

Preface

Like the other titles in the *Learning about Language* series, this volume is conceived equally as a workbook and as a textbook. In most of the chapters the text is interrupted from time to time and you are asked to carry out a particular task. A line has been ruled at the point where these tasks occur. It is important that you take time to complete these tasks; the text will then usually continue with a discussion of the solution to the task. A number of the tasks in this book involve consulting a dictionary. If you do not already possess one, you are advised to acquire a dictionary no smaller than one of the 'concise' dictionaries, e.g. one of the following:

> Longman Dictionary of the English Language
> Longman Concise English Dictionary
> Collins English Dictionary
> New Collins Concise English Dictionary
> Chambers Twentieth Century Dictionary
> Webster's New Collegiate Dictionary
> Concise Oxford Dictionary

Smaller dictionaries cannot be guaranteed to contain the range of information about words that we want to discuss. While this book is not exclusively about dictionaries and the information about words which they contain, it does relate the more general discussion of how we may describe the meaning of words to the ways in which dictionaries do it; and we shall consider in some detail the past, present and future of dictionaries.

Each chapter ends with a number of exercises, practising the points dealt with in the chapter. Like the tasks within chapters, these frequently involve consulting a dictionary. A Key to the Exercises is provided at the end of the book.

I am grateful to the editors of the series, Geoffrey Leech and Mick Short, for their many helpful comments and suggestions on the draft of this book. Any inaccuracies and infelicities that remain are my responsibility.

Acknowledgements

We are grateful to the following for permission to reproduce copyright material:

William Collins Sons & Co. Ltd for extracts from *Collins English Dictionary* 1979, *New Collins Concise English Dictionary*, *Collins/Klett English–German Dictionary*; Victor Gollancz/Doubleday & Co. Inc. (a division of Bantam, Doubleday, Dell Pubg Group Inc.) for the story 'The Princess and the Pea', from *Hans Christian Anderson: The Complete Fairy Tales and Stories* trans. by Erik Christian Haugaard, copyright © 1974 by Erik Christian Haugaard; Longman Group UK Ltd for extracts from *Longman Concise English Dictionary* 1985.

CHAPTER 1

What is a Word?

This book is about words and their meanings. Before we begin to discuss meanings, we need to be clear what we understand by the term **word**. It is an ambiguous term and we use it in many ways, even in ordinary language. If we want to use it as a term in the description of language, we must be sure what we mean by it. To illustrate what I mean by saying that the term 'word' is ambiguous, let me ask you to count the words in the following sentence:

[1] You can't tie a bow with the rope in the bow of a boat.

Probably the most straightforward answer to the question is to say that there are fourteen words in [1]. There are thirteen spaces between the items, and, in writing at least, a word is often regarded as an item bounded by spaces. But the item *can't* is a problem under such a definition, since it is in a sense a coalescence of two 'words', *can* and *not*: part of the abbreviation is recognised in writing by the apostrophe. If we regard *can't* as two words written together (*can not*) and abbreviated, our total now comes to fifteen.

But some of the words occur more than once: *a* and *the*, for example. Are *a* before *bow* and *a* before *boat* to be regarded as (two instances of) the 'same' word and therefore only counted once? Or are they two words, as our counting has so far assumed? And if the two occurrences of *a* and *the* are to be counted as single instances of these words (giving us a total now of thirteen words), what are we to say about the two occurrences of *bow*? As far as the marks on the page (the writing) are concerned, we are dealing with the same sequence of letters: *b* + *o* + *w*. Orthographically, therefore, the two occurrences of *bow* constitute a single word (bringing our total now to twelve). The orthographic perspective taken by itself, of course, ignores the meaning of the words, and as soon as we invoke meanings we are talking about different words *bow*.

What I hope to have shown is that the answer to the apparently simple instruction, 'Count the words in the following sentence', is not simple. You first of all have to ask: 'What do you mean by "word"?' In this chapter we are going to investigate some of the distinctions that need to be made in order to interpret the term 'word' in any particular context.

Written and spoken words

Field linguists investigating a language that does not have a written form often have a problem in deciding where the boundaries between words occur in speech. Written languages have institutionalised word boundaries by means of the orthographic space between words, though, as we shall see, orthographic practice is not always a good reflection of grammar and meaning. And even orthographically there is the occasional difficult case, like *can't* in [1]. To illustrate this point, write an orthographic version of the following representation of English speech, which reflects the fact that there are no 'spaces' between words in speech and that we run words together when we speak:

[2] isediwannidapynamilk.

For those of you familiar with the symbols of the International Phonetic Alphabet (see Knowles, 1987), this would be:

[2a] /ɪsɛdɪwɒnɪdəpaɪnəmɪlk /.

The orthographic version of this is:

[2b] He said he wanted a pint of milk.

But if we did not already have the well established conventions of English orthography, we should have difficulty in deciding, for example, whether *apynamilk* should constitute one word or two, or three or four. In practice, field linguists use a variety of criteria from several linguistic levels – phonological (the level of sounds and their combination), morphological (the level of word structure), semantic (the level of meaning), syntactic (the level of sentence structure) – to decide on word boundaries in languages that they are 'reducing to writing'.

In English, the question of word boundaries in writing still exists in a few cases, especially in how we write **compound words**. Compounds are words that form a unit made up of two or more single words. e.g. *time + keeper, time + lag, time + sharing.* Write down how you would write these combinations.

Nearly all English users would probably write *time + keeper* as *timekeeper.* There may, however, be a difference of opinion with *time + lag.* The possibilities are: *timelag, time-lag, time lag. Collins English Dictionary* (1979), for example, has *time-lag,* while *Webster's New Collegiate Dictionary* has *time lag.* And for *time + sharing* the single unhyphenated orthographic word is probably not found, so that the possibilities are: *time-sharing* and *time sharing.* Check in any dictionaries available to you on how the dictionary recommends that these compounds should be written. In the course of their acceptance into the language as single words, many compounds undergo a development from being written as two words, through being hyphenated, to being written as a single word.

Assuming that we agree, as we do for most words in English, how we should relate spoken and written forms, there still remains for a number of words some confusion between writing and speech. One source of confusion is where a written word may be pronounced in more than one way (by the same speaker!). We have already noted in [1] that *bow* may be pronounced either /baʊ/ and refer to part of a boat, or /bəʊ/ and refer to the result of tying string or a ribbon in a particular way. What are the two different pronunciations and meanings of *sow* and *refuse*?

The same differences of pronunciation apply to *sow* as to *bow*: pronounced /saʊ/, *sow* refers to a female pig; pronounced /səʊ/, it refers to the activity of putting seeds into the ground. When *refuse* is pronounced /rɪˈfjuːz/ it refers to the action of declining or resisting (i.e. it is a verb); pronounced /ˈrɛfjuːs/, it refers to rubbish (i.e. it is a noun).

A more frequent confusion than that of different pronunciations for the same spelling is the reverse: different spellings for the same pronunciation. Look at the following pairs and determine the different meanings of each member of the pair:

[3] feet   – feat
    lesson   – lessen
    fete   – fate
    practice – practise.

You can check the meanings of these items by looking them up in a dictionary.

A particular spelling which has two pronunciations with different meanings represents two different words: there are two words *bow*, and you will find that dictionaries give them separate entries. (We are discounting differences of accent or variation in the pronunciation of words like *either*.) Similarly, a particular pronunciation that has two spellings with different meanings (and we discount variant spellings of words like *medieval/mediaeval*) represents two different words: this is easier to accept, since they are in any case separate headwords in the dictionary, given that the dictionary is based on spelling.

Words which are spelt the same, but have different pronunciations and meanings, are called **homographs**, e.g. *bow*. Words which are pronounced the same, but have different spellings and meanings, are called **homophones**, e.g. *feet/feat*. Additionally, there are many cases where a single spelling and pronunciation occurs with more than one meaning, e.g. *bank*. There are several different 'words' – in terms of items with separate meanings – which are spelt *bank* and pronounced /bæŋk/ in English, e.g.

[4] bank 1 – financial institution
    bank 2 – side of river or stream
    bank 3 – a row of keys on a keyboard.

Words like *bank*, which are spelt and pronounced the same, but have clearly different meanings, are called **homonyms**.

Now consider the following words and say which of them you think are homonyms:

[5] stick banana grow file break.

The clear cases of homonyms in this list are: *stick* (1 = piece of wood, 2 = (cause to) adhere), *file* (1 = proceed in order, 2 = abrasive tool, 3 = collection of papers, 4 = put away in order), and *break* (1 = (cause to) become more than one piece, 2 = vacation, short cessation from work). *Banana* has one clear meaning, referring to a kind of fruit. *Grow* has more than one 'meaning', distinguishable in sentences such as:

[6] They grow a lot of apples in this part of the country.

He's growing a beard.

Don't children's feet grow quickly?

It is questionable, however, whether we can talk of different 'meanings' in the case of *grow*, but rather of variants of a single meaning. We shall consider *grow*, therefore, to be a single word with a number of **senses**, i.e variant, closely related meanings. Such words – and this includes many words in common usage – are said to be polysemous; they are cases of **polysemy** (or multiple meaning). Homonymy, then, refers to words with different meanings sharing the same form (e.g. bank), while polysemy refers to one word having a number of senses or variants of a single meaning.

Now, homonymy and polysemy are not always clearly distinguishable: the variation of a single meaning shades into the recognition of distinct meanings. The senses of *grow* illustrated in [6] are clearly a case of polysemy: they all relate to the meaning of 'development' or 'production'. Further along the continuum towards distinct meanings we might place cases of metaphorical extension, such as *leg* used of a table or *foot* of a mountain: there is probably sufficient continuity of meaning with the literal reference for us to count these as cases of polysemy. The continuity of meaning is, perhaps, a little harder to detect in the case of *eye* when used of a needle, or of *iron* when used of the appliance for pressing clothes: here polysemy is verging towards homonymy.

Note: the terms we have been using in this section are derived from Greek: *homophone, homograph, homonym, polysemy*. *Homo* means 'the same', so *homograph* means 'the same letters', *homophone* 'the same sounds', and *homonym* 'the same name'. *Polysemy* means 'multiple meaning'.

Word-forms

Cutting across and extending our discussion of the spelling and pronunciation of words, we now confront another version of the question, 'Is it the same or is it a different word?' The question also refers to spelling and pronunciation, but it goes beyond just that.

We mentioned earlier that some words have variant spellings: *medieval* and *mediaeval*, for example. This is not just a difference between the spelling conventions of American English and those of British English, but a genuine variation within one standard. American English usually uses *medieval*, but in British English some writers use one spelling and some writers use the other. A similar variation occurs with *foetal* and *fetal*, *aesthetic* and *esthetic*, *judgment* and *judgement*, *gaol* and *jail*, *spelled* and *spelt*, *analyse* and *analyze*: and even more so in the area of proper names, e.g. *Catherine*, *Katherine* and *Kathryn*. In each case we would want to say that these are variant forms of the same word, and not different words.

Spelling is more or less standardised for a particular national variety of the English language; and we can, therefore, speak of genuine spelling variants. It is much more difficult to speak of genuine pronunciation variants, since most variation of this kind can be attributed to differences of regional accent, e.g. the long vowel /ɑː/ and the short vowel /æ/ pronunciations of *grass, path, fast*, etc., associated with southern vs. northern British English accents respectively. In variation due to accent, then, we are talking about systematic variation of sounds, not idiosyncratic variation in the pronunciation of particular words; we would have to look for this within a regional accent or in a supra-regional prestigious accent such as Received Pronunciation (RP), the accent associated with public schools and the BBC. Such variation does occur, e.g. in the RP pronunciation of *either* (/aɪðə/ vs. /iːðə/), *garage* (/gə'rɑʒ/ vs. /'gærɪdʒ/), *economics* (/iːkənɒmɪks/ vs. /ɛkənɒmɪks/). Again we shall want to say that these are variant forms of the same word, and not different words.

More consequential for the notion of 'word' than either spelling or pronunciation variants is variation of word-forms like the following:

[7] girl   girl's   girls   girls'
[8] tiny   tinier   tiniest
[9] sew   sews   sewing   sewed   sewn
Can you explain what kind of variation is occurring here?

The examples in [7], [8] and [9] illustrate the **inflections** found in English for various classes of words: nouns in [7], adjectives in [8] and verbs in [9]. The inflectional forms of the noun in [7] are, respectively, the singular common form (*girl*), the singular possessive/genitive form (*girl's*), the plural common form (*girls*), and the plural possessive/genitive form (*girls'*). Note that the last three of these forms are not distinguished in their pronunciation, only in their spelling. The inflectional forms of the adjective in [8] are, respectively, the base form (*tiny*), the comparative form (*tinier*), and the superlative form (*tiniest*). The inflectional forms of the verb in [9] are, respectively, the base/present tense form (*sew*), the third person singular present tense form (*sews*), the present participle form (*sewing*), the past tense form (*sewed*), and the past participle form (sew*n*). Most verbs in English have the same form for past tense and past participle, e.g. *walked, asked*. (Detailed discussion on inflectional forms can be found in Mugdan, in preparation).

The question that we have to ask is whether *sew, sews, sewing, sewed* and *sewn* are different words or the same word. The answer is not, in fact, one or the other, but yes to both parts of the question: in some sense they are different words, in another they are the same word. Clearly, they are different orthographic and phonological words; they have a distinct spelling and pronunciation (as whole words), even though they have some letters and sounds in common (*sew*, /səʊ/). But the spelling and pronunciation are a reflection of what are essentially grammatical differences. Grammatically, they are different words: they occur in diffrent grammatical contexts. For example, *sews* occurs when the tense is present and the grammatical subject of the verb is third person singular; *sewn* occurs, for instance, in perfect tenses, formed with *have* (*has sewn, have sewn, had sewn*). (Detailed discussion of English grammar can be found in Burton-Roberts, 1986.)

However, as far as the essential meaning is concerned,

sew, sews, etc. can be regarded as the same word. But it is
not a case of polysemy: the dictionary does not enter
different 'senses' for these items. It is rather a case of variant
forms, differing according to grammatical function and
context. The different inflectional forms of *sew* do not refer
to different kinds of activity, nor the forms of *girl* to
different kinds of persons or things. They mark categories
that are considered to be part of grammar: e.g. past tense,
or plural number. Or they reflect a particular grammatical
context, e.g. *sews* occurs when it has a grammatical subject
like *he, she* or *it*.

We have now identified four kinds of 'word'. We need
to summarise and label them. First of all, we have identified
orthographic words, words distinguished from each other
by their spelling. Secondly, we have identified **phonological words**, distinguished from each other by their pronunciation. Thirdly, we have identified **word-forms**, which
are grammatical variants. And fourthly, we have identified
words as 'items of meaning', the headwords of dictionary
entries, which are called **lexemes**. In many cases the item
will be the same for all four kinds of word; e.g. *same* is
always spelt and pronounced in the same way (for a
particular group of speakers), has no grammatical variants,
and is a single lexeme (i.e. is not a homonym). But, as we
have seen, many other cases point to the fact that the sets
of lexemes, word forms, phonological words and orthographic words are not identical. Therefore, the answer to
the question, 'How many words are there in the following
sentence?', depends on what kind of word you are talking
about. Consider [1] again and give an answer to the question for each of the four kinds of word.

[1] You can't tie a bow with the rope in the bow of
a boat.

There are eleven different orthographic words, with two
instances each of *a, the* and *bow*. There are twelve different
phonological words, with two instances each of /ə/ and
/ðə/. There are thirteen different word-forms, since grammatically we must count *can* and *not* as distinct word-forms;
and there are correspondingly thirteen different lexemes.

Lexemes

Earlier we identified 'lexemes' with the headwords of dictionary entries, which implies separate entries for homographs and homonyms (and, since the dictionary is based on spelling, for homophones as well, e.g. *break/brake*), but not separate entries for word-forms. Look at the dictionary entries for *sing, foot* and *bad*, which are verb, noun and adjective, respectively. Which of the word-forms for each of these words (e.g. *sing, sings, sang, sung*) is put in bold type at the beginning of the entry as the **headword**?

The headwords are in each case what are regarded as the **base** forms of the words, from which other word forms are considered to be derived. For verbs, e.g. *sing*, the base form is the present tense form (not third person singular); or alternatively it may be considered as the infinitive form (without *to*). For the verb *be* it is only the latter, since *be* does not figure as a present tense form. For most verbs (e.g. *ask*), all other forms are derived from the base by the addition of suffixes (*asks, asking, asked*), so the base form is the one that is not suffixed. In the case of 'irregular' verbs like *sing*, the same form is considered to be the base (rather than *sang* or *sung*) as for 'regular' verbs like *ask*. For nouns, e.g. *foot*, the base form is the singular common case form: it is singular rather than plural (*feet*) and it is common case rather than genitive/possessive case (*foot's*). For adjectives, e.g. *bad*, the base form is the so-called 'absolute' form (as against the comparative form *worse*, or the superlative form *worst*). For other word classes, e.g. adverb or preposition, where there are no grammatical variants, there is only one form that can be the headword.

   These base forms of words, the headwords of dictionary entries, may be termed the **citation forms** of lexemes. When we want to talk about the lexeme *sing*, then the form that we cite (i.e. 'quote') is the base form – as I have just done – and that is taken to include all the grammatical variants (*sings, singing, sang, sung*).

   What then do dictionaries do about the word-forms of lexemes? Do they account for them in any way? Look again at the entries in your dictionary (I am assuming a desk-size dictionary rather than a pocket-size one, e.g. *Collins English Dictionary, Longman Concise English Dictionary, Chambers Twentieth Century Dictionary*) for *sing, foot* and *bad*, and compare them with the entries for *talk, hand* and *small*. What do you find?

In most dictionaries, if the word-forms can be derived from the base form by the general rules of grammar (inflectional morphology), i.e. the forms are 'regular', then the dictionary entry does not indicate them. This is the case, for example, with *talk, hand* and *small*. If, on the other hand, the word-forms are 'irregular' in their formation, and cannot be deduced from the general rules of grammar, then dictionary entries usually indicate them, as in the case of *sing, foot* and *bad*. A few dictionaries, e.g. *Webster's Third New International Dictionary*, gives the word-forms of all lexemes, whether formed regularly or irregularly, but this is unusual.

We have identified the lexemes of a language with the headwords of a dictionary. This is not entirely correct, at least for most dictionaries, since the entry under a particular headword may contain not just alternative senses (if the lexeme is polysemous), but also any derived lexemes which have been formed from the headword by a process of lexical derivation (suffixation, prefixation, compounding – see further in Chapter 2, p. 30). For example, the lexeme *singer* may be found under the headword *sing*; and *handful* and *handy* appear under *hand* in some dictionaries. These must be regarded as separate lexemes, since, although they are related in meaning to the headword, they usually belong to a different word-class and so are used in different ways in the structure of sentences. Dictionaries, however, differ in the ways in which they treat derived lexemes. Look at the introductory pages of your dictionary to see how it treats these derivations and see if it is consistent in its practice. Check on *choral* (derived from *choir* but with a rather different spelling), and *signify* (derived from *sign* but with a different pronunciation). Does your dictionary treat derivation by prefixes (e.g. *be-friend, en-close*) differently from suffixation (e.g. *friend-ly, clos(e)-ure*)?

If you can, it is worth.looking at more than one dictionary (from different publishers) to see if their practices are different. In most dictionaries you will find the prefixed words at their appropriate place in the alphabetical order as separate entries, e.g. *befriend* under 'B' between *beforehand* and *befuddle*. Suffixed words, on the other hand, are often included within the entry for the word to which the suffix has been added, e.g. *friendly* is included in the entry for *friend*.

The other main subdivisions in dictionary entries (see further in Chapter 3, p. 41) are for the different 'senses' of polysemous lexemes, which are variant meanings of a lexeme. Such is the nature of meaning, that dictionary compilers do not always agree, especially for cases of multiple polysemy, how many senses or indeed which senses to identify for a particular lexeme. Look at the entries in more than one dictionary for the lexemes *think, high, on* and *frog*, and compare the division into senses for each lexeme.

Multi-word lexemes

We have assumed so far that a lexeme consists of no more than one orthographic word. We are now going to revise that assumption and consider sequences of words which have to be considered as single lexemes. We begin with a set of verbs, of which the following are representative:

[10] give in   pass out   think up   put off

In each case the lexeme consists of a verb, followed by an adverb particle. Such combinations are called **phrasal verbs**. When we hear or read such items we understand them as a single semantic unit; in many cases they have a single (usually more formal) verb word equivalent, e.g *succumb, faint, devise, postpone* respectively for the phrasal verbs in [10]. Grammatically, they have the same function in sentences as single–word verbs, except that the adverb particle may be detached from the verb word; compare [11] and [12]:

[11]   Jane has **invented** a good excuse.
[12a] Jane has **thought up** a good excuse.
[12b] Jane has **thought** a good excuse **up**.

Now construct similar sentences to [12a] and [12b] for *put off*, and also substitute a pronoun (*it*) in place of the noun phrase in the position of *a good excuse*. What possibilities of word order do you find?

The possibilities are as follows:

[13] Jane has **put off** the party.
[14] Jane has **put** the party **off**.
[15] Jane has **put** it **off**.

but not

[16] *Jane has **put off** it.

(*Note*: the '*' is the conventional symbol to mean that the sentence is ungrammatical or unacceptable.)

There is another, different, set of verbs that look very similar to the phrasal verbs, of which the following are representative:

[17] look after   think about   speak with   wait for

Try forming sentences equivalent to [13] to [16] with one of these verbs, to determine which are possible in English.

Let us take *look after* as our test verb:

[18] Jane **looks after** her elderly mother.
[19] *Jane **looks** her elderly mother **after**.
[20] *Jane **looks** her **after**.
[21] Jane **looks after** her.

Look after shares the structure of [18] with *put off* [13], but *look after* does not permit the postponement of the particle, as in [14] and [15] with *put off*. *Look after* does, however, allow the particle before a pronoun, in [21], which is not permitted with *put off* [16]. Indeed, the *after* seems to have to remain with the verb word under all circumstances, though there are exceptions in the case of some rather formal constructions, such as:

[22] **After** whom is Jane **looking**?
[23] . . . the proposal, **about** which he is **thinking**

The verbs in [17] are called **prepositional verbs**, and the accompanying particle is a preposition. Clearly, the syntactic operation of prepositional verbs is different from that of phrasal verbs, as [18] to [21] demonstrate. More fundamentally, it has been proposed (e.g. in R. Quirk & S. Greenbaum, 1973, para. 12.5) that the preposition of prepositional verbs can be considered as belonging not to the verb but to the following noun phrase, so that [18] is analysed as: *Jane* (subject) – *looks* (verb) – *after her elderly mother* (adverbial). Against this proposal it can be argued that the preposition is unique to this verb with this particular meaning (i.e. 'care for'/'tend'), unlike the preposition in, for example:

[24] Jane looked **out of** the window.

which could be substituted by other prepositions (e.g. *through, in, over*) and clearly belongs with *the window*. The *after* of *look after* thus belongs at least as much with the verb as with the following object.

The same arguments cannot apply to phrasal verbs, firstly because some phrasal verbs do not take an object in any case (e.g. *break in, walk out*), and secondly because the fact that the adverb particle can be positioned after the object would suggest that it is independent of it. For such reasons, phrasal verbs are usually regarded as single lexemes, whereas prepositional verbs frequently are not, but are regarded as two lexemes, verb and preposition. This difference is reflected in the way that some dictionaries treat phrasal and prepositional verbs. Look up the items in [10] and [17] in your dictionary to see how they are treated, and then if you have it available, look them up in the *Collins English Dictionary (CED)*.

Phrasal verbs, those in [10], are often treated as separate headwords, e.g. in *LDOCE*; whereas prepositional verbs, such as those in [17], are more usually treated as senses or derivations of the verb word, but that may depend on how the meaning of the prepositional verb relates to that of the verb word without a preposition. For example, *look at* is quite similar in meaning to *look*, and may be entered as a 'sense' of *look*: whereas *look after* ('take care of') is rather different in meaning and we would not expect to find it as a 'sense' of *look*.

Another case of multi-word lexemes is that of some compounds. We noted earlier that orthographically the word boundaries in respect of compounds are indeterminate. Some compounds are written as one word (*timekeeper*), some as one word but hyphenated (*time-consuming*), and others as two distinct words (*time machine*). Compounds are clearly to be regarded as single lexemes, and cases like *fire extinguisher* must therefore be considered multi-word lexemes. There are many compounds of this type, and they are continually being coined, e.g. *child safety seat, manhole cover, rear-view mirror*. However, structures of this kind are not always compound lexemes: there appears to be a gradation between compounds (regarded as single lexemes) and syntactic constructions (of several lexemes) which look like compounds, e.g. headlines like 'City business racket investigation'.

Any newspaper or magazine will contain a number of compounds. You could take a newspaper and make a list, and then see if your dictionary contains them as single lexemes, either as separate headwords (unusual) or as derivatives under one of the constituents.

A final kind of multi-word lexeme is the idiomatic phrase, a more or less fixed sequence of words with a unitary meaning, such as:

[25]   hand in glove
       spill the beans
       let the cat out of the bag
       get the wrong end of the stick.

These phrases clearly consist of more than one orthographic or phonological word, but they each have a unitary meaning and sometimes have a single-word equivalent; e.g. *spill the beans* = *reveal* (a secret), *get the wrong end of the stick* = *misunderstand*. The characteristic of an idiom (see further in Chapter 7, p. 106) is that its meaning is not the sum of the meanings of its constituent parts; it is to be interpreted non-literally, as a whole. It is a single lexeme. Dictionaries, too, treat idioms as single lexemes; they are usually entered as a derivative under one or more of the constituent words of the idiom, presumably the one(s) that are considered to be central to the idiom. For example, in *LDOCE, hand in glove* is entered under *hand, let the cat out of the bag* under *cat, spill the beans* under both *spill* and *bean, get the wrong end of the stick* under *wrong, end* and *stick.* You may like to check in your dictionary on how it deals with these idioms.

Lexical and grammatical words

There is one more distinction that it is useful to draw when discussing the notion of 'word', and it is one which cuts across all the other distinctions that we have made so far. Let me ask you, first of all, to rewrite the following sentence as if it were a telegram, and to note which words you leave out:

[26] I'm coming tomorrow on the train at six o'clock.

The telegram equivalent of [26] would be something like:

[27] Coming tomorrow six o'clock train.

The words that are omitted are: *I, am, on, the, at.* You will notice that they are all very short words, and that they are not essential to the basic interpretation of the sentence. However, they do help to make the sentence's meaning explicit, and they would be essential if [26] were part of a letter, for example. The function of the omitted words, though, is rather different from that of the telegram words of [27]: the necessary words for the telegram bear the main burden of referential meaning (see further in Chapter 4), while the omitted words make the sentence grammatically complete and provide relations to other sentences within a text. They may be regarded as the bricks and mortar of sentences, respectively. The bricks, the words included in [27], are often called **lexical** words; and the mortar, the words of [26] omitted from [27], are called **grammatical** words, or 'function' words.

Lexical words belong to classes (or subclasses) of words which are relatively large and open, viz. nouns, most verbs, adjectives, many adverbs. There are some subclasses of verb (e.g. *am* in [26]) and adverb (e.g. *now, then*) which are more like grammatical words than lexical words. The lexical classes are open in the sense that their membership is not stable; new items are continually being coined, and some become obsolete and fall out of use. Grammatical word-classes, by contrast, have a relatively small and stable membership; they include pronouns, determiners (words that accompany nouns and 'determine' their contextual status, e.g. *the, a, this, my*), prepositions, conjunctions, auxiliary verbs, some adverbs. The membership of these classes changes only very slowly over time. For example, the subclass of pronouns called personal pronouns, whose membership (in standard English) now comprises *I, me, you, he, him, she, her, it, we, us, they, them*, has not changed for over three hundred years, following the loss of the *thou* and *thee* forms, which are found in Shakespeare and the King James version of the Bible (1611). However, the *thou* and *thee* forms continue in some dialects, while other dialects have introduced a plural 'you' (*yous*), and some varieties of American English have *you-all* as the plural of *you*.

Look now at the following sentence and say for each word whether it is a lexical word or a grammatical word:

[28] My aunt has given up going there frequently, because the food is so bad.

The clear members of the lexical word-classes in [28] are: *aunt* (noun), *given up* (single-lexeme phrasal verb), *going* (verb), *frequently* (adverb), *food* (noun), *bad* (adjective). The remaining words have a more or less grammatical function. *My* belongs to a restricted set of possessive determiners; *the* belongs to an even more restricted subclass of determiners, viz. the articles, which, besides *the*, includes only *a/an*. *Has* belongs to the subclass of auxiliary verbs, used to form certain tenses of the verb. *Is*, a form of the verb *be*, has here the function of a copular (= joining) verb; it is not so clearly grammatical as the auxiliary verb *have*, but neither is it so lexical as the verb *give up*. *Because* belongs to the class of conjunctions, which like the class of prepositions is fairly numerous though closed in membership, and could be regarded as falling somewhere midway between grammatical and lexical. *There* and *so* belong to restricted subclasses of adverb: *there* is a kind of pro-adverb (like a pro-noun), which stands for an adverbial expression occurring previously in a text; *so* is a modifying adverb that 'intensifies' an adjective or other adverb. We might mention also that the *up* of *given up* can be considered the member of a grammatical subclass of adverb particles, which, when not used as a constituent of a phrasal verb, often have a pro-adverb function, e.g. in:

[29] She's gone out.

As I have indicated in the discussion above, the distinction between lexical and grammatical word-classes should not be drawn too rigorously. There is probably a gradation between completely lexical (e.g. nouns) and completely grammatical (e.g. articles), with many classes falling somewhere between these two extreme points, as the following diagram illustrates.

NOUN	PREPOSITION	PRONOUN	DETERMINER
VERB	CONJUNCTION	ADVERBS like	(e.g. *the, this*)
ADJECTIVE	QUANTIFIER	*here, now*	AUXILIARY VERB
ADVERBS in -*ly*	ADVERBS like	POSSESSIVE	
	however	DETERMINER	
		(e.g. *my*)	

Most Lexical ⟵——————————————————⟶ Least Lexical
Least Grammatical ⟵——————————————————⟶ Most Grammatical

We began this chapter with a question about the meaning of the term 'word'. We have shown that the term is used in a number of related ways in different contexts, and we have drawn several distinctions relevant to a discussion of words and their meaning. We have cleared the ground for our subsequent investigations.

Exercises

1. Say which of the following may be considered to have or to be homographs, homophones, homonyms or polysemy:
   sea   break   line   ear   prayer   mature   trace   house
2. List the word-forms (grammatical variants) of the following lexemes:
   child   run   little   fly   basic   turn
3. How many different lexemes, word-forms and orthographical words are there in the following sentence?

At the second drum roll they have to roll out the flag
and have it up the mast in fifteen seconds.

4. Which of the following would you regard as multi-word
   lexemes?
   take care of   *look into*   *browse among*   *story book*
   garden fence *send off for* *over the moon* *training weekend*
   fierce tiger *look up*

5. Identify the 'lexical' and the 'grammatical' words in the
   following:

   When I am grown to man's estate
   I shall be very proud and great,
   And tell the other girls and boys
   Not to meddle with my toys.
   (R. L. Stevenson: 'Looking Forward')

CHAPTER 2

Where Did English Words Come From?

The word 'English' comes from the name of one of the three Germanic tribes which invaded and settled in this island during the fifth and sixth centuries: the Angles. The Angles settled in areas of England to the north of the River Thames, whereas the Jutes, who were the first of the tribes to arrive, settled in what is now Kent, the southern part of Hampshire and the Isle of Wight. The remaining areas of southern England were settled by the Saxons. The three tribes probably spoke mutually intelligible dialects, and the language of the country as a whole seems to have been known as 'Englisc' from this period. We, however, often refer to this early form of English as 'Anglo-Saxon' and to the words that originate from this period as the 'Anglo-Saxon' words of the English vocabulary. The time from the middle of the fifth century, when the settlement by the Anglo-Saxon tribes began, until the end of the eleventh century, following the Norman French Invasion, is traditionally called the 'Old English' period of the language.

Origins

English, in common with most of the languages of Europe (with the notable exceptions of Finnish, Hungarian and Basque) and of North India, is considered to belong to the Indo-European family of languages. Proto-Indo-European is the supposed parent of all these languages, spoken some time before 3000 BC. It was 'reconstructed' by nineteenth-century historical linguists on the basis of the comparison of the oldest known languages for which written records had survived; e.g. Latin, Greek and Sanskrit (an ancient language of North India). One of the 'branches' of the Indo-European family tree is Germanic, and Proto- or Primitive Germanic, another reconstructed language, is said to be the ancestor of the Germanic languages now spoken in central and northern Europe. It, in turn, had three branches: North Germanic, East Germanic and West Germanic, though whether these terms ever corresponded to actual languages is uncertain.

The North Germanic branch developed into today's Scandinavian languages, notably Danish, Swedish, Norwegian and Icelandic. The East Germanic branch died out; its only known member was Gothic, from which there survives a Bible translation by a Bishop Wulfila in the fourth century. The West Germanic branch developed into modern German, Dutch, Frisian and English. English's closest relative is, in fact, Frisian, spoken in the north-west Netherlands and the islands nearby, known as Friesland.

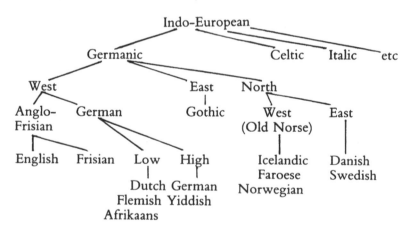

Diagram showing relationships between languages in the Germanic branch of the Indo-European family.

When the Angles, Saxons and Jutes – West Germanic tribes – came to this island in the fifth century and in increasing numbers in the sixth century, the people that they displaced were Celtic, as were most of the earliest known inhabitants of western Europe. Many Celts fled before the Germanic invaders to the fringes of the country – Wales, Cornwall, the Scottish highlands – or escaped across the sea to their Celtic relatives in Brittany. Those who remained became assimilated to the new tribes by intermarriage. Traces of the Celtic language spoken by the 'Ancient Britons' survive in modern English mainly in the names of rivers (Avon, Dee, Ouse, Severn, Stour, Tees, Thames, Trent, Wye, etc.) and of some towns and cities (London, Dover, Crewe, Carlisle, Leeds, Penrith, York, etc.). The language of the subjugated and dispossessed people left only a small impression on the language of the new inhabitants of what was now being called 'England'.

English is, then, essentially a Germanic language. The vocabulary of modern English, one might assume, originated in the languages of the invading Anglo-Saxon tribes of the fifth and sixth centuries. But look at a page of a dictionary (e.g. *Longman Concise English Dictionary* or *Collins English Dictionary*) and see how many of the words whose origins are given (usually at the end of the entry in square brackets) are marked as deriving from Old English (abbreviated 'OE').

Taking a page at random from the *Longman Concise English Dictionary* (p. 616) with the words from *gratify* to *graze*, we find the following that have their origins indicated:

- two words from Old English: *grave, graze*
- four words from French: *gravel, gravimeter, gravure, gravy*
- four words from Latin via French: *gratify, gratitude, grave* (= 'serious'), *gravity*
- five words directly from Latin: *gratis, gratuitous, gravamen, gravid, gravitas*

Clearly this is a very small sample and is probably not entirely representative of the modern English vocabulary as a whole, but I am sure that you probably found something like this pattern on the page of the English dictionary you looked at. In my sample, two out of the fifteen are given as being of Old English (or Anglo-Saxon) origin: that is a little over thirteen per cent.

This means, then, that a very large number of words have been incorporated into the vocabulary of English from other languages. Such words are often called **loan-words** and the process by which they are brought into the language is called **borrowing**. As we can see from my sample above, important sources for borrowing into English have been Latin and French; but, while English speakers have borrowed heavily from these two languages, they are not the only sources of loan-words for English, as we shall see. When words are borrowed, they frequently become **nativised** in the course of time: the pronunciation and the grammatical inflections, and perhaps even the spelling, are adapted to the system of English; and their foreign origin is obscured. Few English speakers, apart from those interested in the origins of words, would think of *gravel* or *gravy* as loan-words; and most would probably think of *gratify*, *gratitude*, *gravity* and *gratuitous* as native English words. It is only words like *gratis*, *gravitas*, and perhaps *gravure*, which still retain some appearance of 'foreignness'.

Without consulting a dictionary, which of the following words would you say were 'native' English words and which loan-words?

[1] cheese   curtain   drift   empty   flavour   grain
     hepatitis   imprimatur   jar   khaki   limit   meal
     misogynist   nostril   oppress   poverty   quench
     reprove   segment   skin   token   usher

Out of these twenty-two words, seven are indicated in the *Longman Concise English Dictionary* as originating from Old English (*cheese, drift, empty, meal, nostril, quench, token*), and one from Old Norse (*skin*). Of the remainder, a few retain their foreign appearance – *hepatitis, imprimatur* (both from Latin), *misogynist* (from Greek), *khaki* (from Hindi) – and would probably be recognised as foreign by most educated English speakers. Your identification of the other words as originally loan-words probably depends on your knowledge of French and Latin and your ability to recognise similar sounding or similarly spelt words in those languages. They all originate ultimately from Latin, though most of them were borrowed into English from French at various times.

Anglo-Saxon words

Words of Old English origin constitute the bulk of the vocabulary used in everyday conversation. Whereas, then, they are in a minority as far as the vocabulary of modern English as a whole is concerned, they are in the majority in the vocabulary of normal daily conversational intercourse. This is in part explained by the fact that Anglo-Saxon words occur more frequently than loan-words; more instances of *have, take* or *to* will occur in any discourse than, say, *surprise* or *concern*. Anglo-Saxon words tend to be short, consisting of one or two syllables; though many borrowed words, especially those that have been nativised, are also of one or two syllables, as the list at [1] above demonstrates. However, words of three or more syllables are nearly always of foreign origin, even if their foreignness is no longer recognisable.

You can examine the proportions of Anglo-Saxon and loan-words in spoken discourse by tape-recording some ordinary conversation, making a transcription and identifying the origins of the words used.

In identifying the Anglo-Saxon words of modern English as typical of everyday conversation, we are implying that they are characteristically associated with a particular stylistic level, namely that of informality or of colloquial language. Indeed there is an expression in modern English, 'speaking Anglo-Saxon', which means 'plain, blunt speaking'. Alternatively, the expression 'Anglo-Saxon words' is used as a euphemism for taboo words or 'four-letter' words. We associate words of Anglo-Saxon origin with straightforward, direct speaking in face-to-face interaction.

Indeed, English vocabulary contains many pairs of words, one of whose members is of Old English origin and belongs to the register of informality, whereas the other member is a borrowed word and tends to be associated with more formal contexts. If you look up the origins of the following pairs of words in a dictionary, you will see this point illustrated:

[2] pluck   – courage (French)
    sweat   – perspire (French)
    guts   – determination (Latin)
    clothes – attire (French)
    climb   – ascend (Latin)

begin   – commence (French)
book   – volume (French)
pride   – hubris (Greek)
lung   – pulmonary (Latin)

The items in the left-hand column all originate from Old English and tend to be everyday, informal words. Those in the right-hand column are all borrowed words (from French, Latin or Greek) and tend to be associated with the written language or more formal contexts.

Borrowed words

We turn now to consider the sources of borrowed words in English. Words have been borrowed sporadically down through the centuries as English speakers in some numbers came into contact with other cultures and other languages. The excursions of medieval knights on crusades to the Holy Land and the consequent contact with Islamic culture and the Arabic language opened up the way for borrowings into English like *emir, mohair, mufti*. Similarly the Second World War produced its crop of borrowings from German, such as *blitzkrieg, luftwaffe* and *Gestapo*. But there have also been times when large-scale borrowings have been made into English. This was especially the case in the century or so following the Norman Conquest (1066), when the language of the dominant culture in England was French; and again in the sixteenth and seventeenth centuries, when, following the Renaissance, Latin became increasingly important as the language of learning and academic life. We can, then, perhaps distinguish between borrowings that have resulted from incidental cultural contact, those that have resulted from political conquest or invasion, and those that have resulted from cultural 'invasion'.

The Anglo-Saxons had been in possession of the land for some two centuries before the first major political invasion occurred. Beginning in the late eighth century and continuing through into the eleventh century, waves of 'Viking' invaders came from Scandinavia. Many, especially later on, settled and made their home in England, predominantly in the eastern part of the country (e.g. East Anglia, Lincolnshire, Yorkshire, Northumbria), but also in the north-west (especially Cumbria). The invaders, mostly

Danes and Norwegians, spoke dialects of what has been called Old Norse, the parent language of the modern Scandinavian languages. It was a Germanic language of the North Germanic branch, and so was related to the Old English spoken by the Anglo-Saxons.

Unlike the Normans who were to follow them, the Scandinavian invaders never achieved political and cultural dominance over the whole country, although they did rule the north-eastern part of the country, known as 'Danelaw', for a while, and a Danish king (Canute) ruled over all England in the earlier part of the eleventh century. But the Scandinavians eventually became absorbed and assimilated into the native Anglo-Saxon culture. Consequently, their linguistic influence on English was relatively small. In any case, Old English and Old Norse, as sister Germanic languages, had many words in common.

A number of words beginning with *sk* were borrowed from Old Norse at this time, e.g. *skill, skin, skirt, sky*. The *sk* betrays their origin: the equivalent Old English consonant combination, written *sc*, came to be pronounced *sh*, so that, for example, *shirt* from Old English and *skirt* from Old Norse derive from the same source-word. Interestingly, too, the Old English third person plural pronouns (*hie, hiom, hiera*) were replaced with the Scandinavian equivalents giving modern English: *they, them, their*. Perhaps, though, like the Celts, the most enduring legacy from the language of the Scandinavian invaders is to be found in numerous place-names, which are to be found in northern and eastern England. The most widespread relic of Old Norse is probably the ending *-by* in places like *Whitby* and *Derby*, meaning simply 'village'. Other striking endings include: *-thwaite*, as in *Bassenthwaite*, meaning 'clearing'; *-scale*, as in *Seascale*, meaning 'hut'; *-thorp(e)* as *Mablethorpe*, meaning 'small village'.

Look at the following two lists of words. One list contains words which originate from Old Norse, and the other contains words originating from Old English. Can you judge which list is which?

[3a]   anger   get   hit   husband   ill   raise
       scrape   take   ugly   want
[3b]   hate   fetch   strike   bridegroom   sick   rise
       slide   bring   foul   need

The words borrowed into English from Scandinavian (Old Norse) are in the [3a] list, and those of Old English origin are in the [3b] list. In general it is not possible to predict from the shape of the word what its origin might have been (compare *shirt/skirt* above): this is not surprising when we consider the close relationship between Old Norse and Old English.

The second, and linguistically far more significant political invasion was that of the Normans under William the Conqueror in 1066. Not until Henry Bolingbroke acceded to the throne in 1399 would an English monarch have English as his native language. William and his followers spoke the northern French dialect of Normandy, but by the time Normandy was lost to France in 1204 the influence on England was increasingly that of central French. Unlike the Scandinavian invaders, the French formed a cultural, social and political elite, dominating whole areas of public life, like government, the law, the church, and the life of the court and the manor house. These were also the areas – with the addition perhaps of commerce – where writing had an important function, so that for nearly three centuries English fell into comparative disuse in the written medium: French and Latin dominated. Parliament was first opened in English in 1362, and the 'Statute of Pleading' was enacted, which provided for the conduct of law-court proceedings to be in English. In order for government to be carried out, it was necessary for intercourse between governors and governed to take place, and inevitably it was the language of the governed (now in the phase customarily termed 'Middle English') which borrowed large numbers of words from the language of the governors.

As we have noted before, borrowed words of French origin often represent equivalents for native English words on a 'higher', more formal stylistic level; e.g. *child – infant, begin – commence, hearty – cordial, happiness – felicity*. But in some cases the stylistic distinction is apparently absent, e.g. *weapons – arms, thief – robber*. More significant at this period, though, is the wholesale borrowing of French words to refer to French-dominated areas of life, such as the law: *justice, jury, judgement, estate, equity, lease, legacy, libel, perjury*, and many others were borrowed into English during this period. Similarly, words referring to the social and cultural pursuits of the nobility were also borrowed, for example, gastronomic terms like *grill, fry, stew, boil* and *roast*, or terms used in the hunt like *chase, quarry, scent* and *track*, or titles of the nobility like *prince, duke, viscount* and *baron*, or terms of combat and chivalry like *courtly, amiable, favour, lance, generous* and *enemy*.

Words derived from French are sometimes recognisable from their characteristic patterns of spelling, e.g. the *-ity* endings of *felicity* and *equity*, the *-our* ending of *favour*, or the *-ant* of *infant*. Without consulting a dictionary, and using your general knowledge of the medieval period, which of the following words would you judge to have been borrowed from French?

[4] freedom   liberty   amity   friendship   royal   kingly
    strange   odd   lie   perjury   malice   ill-will
    dignity   worth   glass   mirror   sheep   mutton
    gentle   kind

The words in [4] borrowed from French in the medieval period are: *liberty, amity, royal, strange, perjury, malice, dignity, mirror, mutton* and *gentle*. The other words are mainly of Anglo-Saxon (OE) origin, though *odd* originates from Old Norse.

No foreign political invasions of the British Isles have taken place since the Norman Conquest, but we might speak of a 'cultural' invasion consequent on the Renaissance with its revival of interest in the classical languages and cultures of Greece and Rome. Latin had for centuries been the language of the church in western Europe and we might date the first significant 'cultural' invasion of Latin at 597, when Augustine landed in Kent with the commission from Pope Gregory I to convert England to Roman Christianity. It was the Latin form of Christianity which eventually dominated over the Celtic Church which had been established by Irish missionaries in the north of the country. From this time we find a number of ecclesiastical terms borrowed into (Old) English, such as *monk, bishop, priest, abbot, canon* (= 'teaching', 'dogma'), *altar*. Many of the more technical theological terms, however, seem to have been borrowed from Latin during the medieval period, such as *salvation, repentance, resurrection, ascension, eucharist, baptize.*

After the Renaissance, in the fifteenth and sixteenth centuries, Latin became more generally the language of learning, and along with Greek was widely regarded as superior to the vernacular languages of Europe like English. Indeed, it was thought that the nearer the vernacular languages approximated to Latin and Greek, the more 'perfect' they would become. We find, then, a considerable borrowing of Latin and Greek words at this time into the 'learned' vocabulary of English, a process that has continued to the present day, especially in the scientific and medical fields. From Greek, sometimes by way of Latin, we have borrowed words like *agnostic, diagnosis, athlete, catastrophe, encyclopaedia, climax* and many more. From Latin the borrowing has been even more extensive, including words like *arbitrator, explicit, index, major, minor, proviso, simile,* and many more. In modern English many technical words are formed from Latin and Greek **combining forms**, e.g *television, electro-cardio-graph, poly-technic, amino-methane.* Borrowings from Latin and Greek can often be recognised from their typical endings, e.g. Latin *-um* (in *quorum, referendum, symposium*), *-us* (in *campus, chorus, fungus, -a* (in *diploma, drama, formula*), and *ex/-ix* (in *index, appendix, matrix*). In Greek borrowings, typical endings include *-is* (in *analysis, crisis, synopsis*) and *-on* (in *automaton, neutron, phenomenon*).

Some of the following words have been borrowed into English from Greek and some from Latin. Make a guess at their origin before you look them up in a dictionary.

> [5] chromatic   criterion   dithyramb   egregious
>   enthusiasm   homologous   immediate   lethal
>   memorandum   monotone   orchestra
>   promiscuous   scalpel   transmit   vacuum

The following words in [5] have been borrowed from Greek: *chromatic, criterion, dithyramb, enthusiasm, homologous, monotone, orchestra.* The others were borrowed from Latin.

The languages and periods that we have discussed so far are the ones that represent large-scale borrowings into the English vocabulary. There has continued to be borrowing on a smaller scale not only from French and

Latin, but from many other languages as well. In many cases the spelling of a word, if not the pronunciation, betrays its origin. From French we have *avant-garde, chic, detente*; from German *rucksack, kindergarten, alpenstock, ostpolitik*; from Italian we have numerous musical terms like *cantata, concerto, oratorio, opera, soprano*; from Spanish, via the New World, we have *chocolate, cocoa, potato, tobacco, vanilla*; from Russian *pogrom, bolshevik, czar, balalaika, samovar, sputnik*; from Indian languages, as a result of colonisation, we have surprisingly few, but they include *bungalow, dinghy, shampoo, cot, juggernaut*, and some slang expressions like *char* (in 'cup of char') and *wallah*: likewise there are a few from Australian aboriginal languages: *boomerang, budgerigar, kangaroo, wombat.*

Now make a guess at the origins of the following words. Then check your guesses with the discussion below – and perhaps with a dictionary as well.

[6] angst anorak apparatchik barbecue bonanza boutique diktat discotheque flak hinterland intelligentsia jukebox karate ombudsman pyjamas reportage robot tea troika tycoon

---

In [6] the following words have been borrowed into English from French: *boutique, discotheque, reportage*; the following from German: *angst, diktat, flak, hinterland*; the following from Russian: *apparatchik, intelligentsia, troika*; the following from Japanese: *karate, tycoon*; the following from Spanish: *barbecue* (via American Spanish), *bonanza. Anorak* comes from Eskimo; *jukebox* (at least the *juke* part) comes from the Gullah dialect spoken by blacks in North Carolina and Georgia, USA; *ombudsman* comes from Swedish; *pyjamas* from Hindi; *robot* from Czech; and *tea* from the Amoy dialect of Chinese.

Making new words

Borrowing words from other languages is not the only way in which the vocabulary of a language may be expanded. A number of linguistic processes may operate to enable speakers to coin new words from those that are already in the vocabulary. Only rarely is a word created from nothing but the sounds or letters of the language; there is usually some motivation or linguistic process at work, though this is not always the case with product names. There is no obvious motivation for *Daz* or *Persil*, apart from the way they sound; but other products are named after their inventors or manufacturers (*Biro, Hoover*), or are motivated in some way by sound-plus-meaning, e.g. *sellotape (seal + tape), xerox* (from *xerography*). Others are motivated by the associations of the names given, e.g. *Aquafresh* (a brand of toothpaste), *Frish* with its assocation with 'fresh' (a brand of toilet cleaner), *Outline* (a brand of low-fat margarine). Some of these trademarks come into the vocabulary at large to stand for the kind of product in general or the process associated with it: *hoover, sellotape* and *xerox* are used as verbs in modern English (*sellotape* only in British English), e.g.

> [7] hoover the lounge (i.e. clean using a vacuum cleaner) sellotape this parcel (i.e. wrap using adhesive tape) xerox this document (i.e. make a photocopy of)

Some words we do not know the origin of. The 1960s word *hippie*, for example, is a derivation from the adjective *hip*, which is a variant of an older (1940s) form *hep*; but we have no knowledge of where or when or how the word *hep* was coined. The inventors of a few words are known. The seventeenth-century poet John Dryden, for example, is credited with the invention of *witticism*, but the motivation for it is clear – it is a derivation from *wit* (an Anglo-Saxon word) by analogy with *criticism*. For some words, whose origins are otherwise unknown, we assume that they have been coined because their sound imitates the object or action referred to, e.g. *flick* or *ping*. But a great many words are coined by the application of productive linguistic processes, and to these we now turn (see also Mugdan, forthcoming).

One of the most productive ways in which new words have been coined, especially in modern times, is by the process called **compounding.** Compounding involves combining two or more existing words in order to form a third, new, word. For example, the noun *double-glazing* is a compound formed from the adjective *double* and the present participle (verbal noun) *glazing*; the noun *motorway* is formed from the two nouns *motor* and *way*; the noun *gobstopper* is formed from the noun *gob* (a slang word for mouth, borrowed from Gaelic) and the deverbal noun *stopper* (i.e. derived from the verb *stop*). Most compounds are nouns; they are coined because there is a need to 'name' an object or thing that has not been named before. And because the meaning of a compound is usually transparent (i.e. it can be deduced from the meanings of the words from which it is formed), it readily commends itself to acceptance by the speakers of the language. Not all compounds are nouns; indeed most word-classes may contain compounds, e.g. *overcharge* (verb), *lacklustre* (adjective), *outside* (adverb), *into* (preposition), *yourself* (pronoun).

There is another kind of compounding, in which the parts of the compound are not themselves independent words. These are compounds formed from the Latin and Greek loan-words that we mentioned earlier. In a word like *bibliography*, for example, neither *biblio-* nor *-graphy* are words in English, though they are, with suitable inflections, in Latin or Greek, with the meanings 'book' and 'writing' respectively. We refer to these compounds as 'neo-classical' compounds, and to their parts as 'combining forms'. Many scientific and academic words continue to be coined using the combining forms borrowed from Latin and Greek, such as *bio-*, *electro-*, *tele-*, *-ology*, *-phile*, *-scope*.

Another highly productive process by which new words are coined is **derivation**. Derivation involves adding to an existing word either a suffix (at the end) or a prefix (at the beginning). Suffixes and prefixes, known collectively as affixes, may not stand alone as words; they occur only in combination with a word. For example, the noun *location* is (was) derived from the verb *locate* by the addition of the suffix *-ion*; and the negative form *dislocate* (verb) or *dislocation* (noun) is (was) derived by the addition of the prefix *dis-*. To say that a word 'is derived' from another is a descriptive statement about a linguistic process that relates items in the vocabulary; to say that a word 'was derived' from another is to regard the process as having taken place at some time in the history of the language.

Frequently, as we see with *location*, the function of an affix (particularly suffixes) is to derive a related word in a different word-class: *-ion* changes verbs to nouns. Alternatively there is no change of word-class. Sometimes the change is from one kind of word to another kind of word in the same word-class, e.g. *-hood* changes a 'concrete' noun to an 'abstract' noun, as in *childhood, priesthood*. Or the affix (especially prefixes) adds some variant of meaning to the word that is subject to derivation, eg. the negative meaning of *dis-* in *dislocation* or *un-* in *unrepentant*.

What function (word-class change) or meaning (e.g. negative) do the following affixes have? Many dictionaries include derivational affixes among their entries, but rather than consult a dictionary you should try to make your deduction from examples that you can think of.

[8] Suffixes: -ation -ful -ify -ly -ment
    Prefixes: anti- arch- be- en- re-

-ation derives a noun from a verb, e.g. *computation*
-ful derives, usually, an adjective from a noun, e.g. *peaceful*
-ify derives a verb from an adjective, e.g. *purify*, or from a noun, e.g. *horrify, countrify*
-ly derives an adverb from an adjective, e.g. *briefly*; or an adjective meaning 'having the qualities of' from a noun, e.g. *cowardly*
-ment derives a noun from a verb, e.g. *entertainment*
anti- means 'opposite' or 'against', as in *anticlockwise* or *antitheft*
arch- means 'chief' or 'supreme', as in *archbishop* or *archenemy*
be- often derives a verb from an adjective, e.g. *belittle*, or from a noun, e.g. *befriend*
en- derives a verb from a noun, e.g. *enthrone, enslave*, or an adjective, e.g. *enrich*
re- means 'again' or 'anew', e.g. *reprint, rehouse*

A further kind of derivation, but one which does not involve the use of affixes, is the possibility of using a word as a member of a word-class other than the one to which it is normally assigned. There are, for example, in English a number of words which occur both as nouns and as verbs, e.g. *bottle, skin, catch, jump*. It is probable that *bottle* and *skin* were originally nouns, which were subsequently used as verbs; while the reverse is probably the case for *catch* and *jump*. We find a similar process operating with *hoover* and *xerox* in their use as verbs. This derivational process is known as **conversion**: a word is converted from one word-class to another without change of form.

Finally we review a number of interesting but minor word-formation processes. Words have been coined by **blending** two words together and retaining part of each, e.g. *telegenic* is a blend of *television* and *photogenic, permafrost* is a blend of *permanent* and *frost*. Words may be coined by the process of **clipping**, or abbreviation, such as *fridge* from *refrigerator, pram* from *perambulator, exam* from *examination, ad* or *advert* from *advertisement*. Words may be coined by **back formation**, which involves the removal of affixes; e.g. *babysitter* (noun) preceded *babysit* (verb), *double-glaze* (verb) is derived from *double-glazing* (noun). And lastly, words may be **acronyms**; that is, they may be composed of the initial letters of the words of a phrase. This occurs especially with the names of organisations, e.g. *UNESCO* (from: United Nations Educational, Scientific and Cultural Organisation); but a small number of ordinary words are in fact acronyms, e.g. *laser* (from: *l*ight *a*mplification by *s*timulated *e*mission of *r*adiation).

We have seen in this chapter how the words in the vocabulary of modern English originate from a number of sources. Some go back to the language of the Anglo-Saxon invaders, Old English. Many have been borrowed from a variety of languages, especially French and Latin. Others have been formed from words already existing in the vocabulary at a particular time by compounding or derivational processes. The words that have undergone derivation may themselves have originally been loan-words (e.g. *courtly*, with *court* from French, and the *-ly* suffix from Old English), as may the affixes making the derivation (e.g. *roughage* with *rough* from Old English and *-age* from French); or the parts of a compound may be borrowed separately from Latin or Greek as combining forms, sometimes producing hybrids like *television* (Greek + Latin). These are the main ways in which the modern English vocabulary has expanded to its present compass; but the ingenuity of speakers to coin new forms seems unbounded. You have to convince your fellow-speakers, though, that the word is worth adding to the vocabulary.

Exercises

1. Which six words from the following list do you think originate from Anglo-Saxon? Remember that words

derived from Anglo-Saxon tend to be short and to be associated with the daily goings-on of ordinary life.

let   letter   lettuce   lever   lewd   liar   libel   library   lick   lid   life   ligament

2. Which eight words from the following list do you think were borrowed from French? Remember that French words sometimes betray their origin by their spelling and are often associated with a more formal style of language.

pedal   pedometer   peg   peignoir   pellet   pencil   penny   pension   pepper   perform   perfume   pin

3. Which words from the following list do you think were borrowed directly from Latin?

subdivide   subsidy   suburb   such   suck   suction   suede   suffix   sugar   suggest   sun   superb

4. Supply a 'formal' French or Latin loan-word equivalent for each of the following 'informal' Anglo-Saxon words.

cheap   cheeky   hard   lighting   busy   buy   worker   cross (verb)   *own* (verb)   *give*

5. Form as many neo-classical compounds as possible with the initial combining form *geo-* (= 'earth'). *Note*: *geography* and *geometry* are not neo-classical compounds; they existed as whole words in Latin/Greek, and *geometry* was borrowed into Middle English.

6. Try forming new derivations as indicated below. Could you convince your fellow-speakers that they are worth adding to the language? E.g. derive a noun from the verb *forget*. You might suggest *forgettance* (by analogy with *remembrance*), as in 'Every forgettance of her consoles him'.

   (a) derive nouns from the verbs: *table* (e.g. a proposal), *defy*

   (b) derive adjectives from the nouns: *minute* (time), *widow*

   (c) derive nouns from the adjectives: *worse, see-through*

CHAPTER 3

Dictionaries: the Repositories of Words

Together with a copy of the Bible, a dictionary must be the most likely book to be found in the majority of English homes. And like 'the' Bible it is often referred to as 'the' dictionary, as if it were a version of a single common book. In some senses it is: lexicographers work in a tradition (see Chapter 8) that has defined what a dictionary should look like; there is considerable commonality between dictionaries of the same size; and there is even unacknowledged 'borrowing' from one dictionary to another, if only in the matter of the choice of which words to include. Many people think of 'the' dictionary as representing 'the English language', making the assumption that language is about words and 'the dictionary' is the collection – the repository – of the words of the language. Consequently 'the dictionary' is viewed as having authority in matters of language usage of all kinds (we shall consider some of the users of dictionaries in Chapter 13). As we shall see later on, it is not justifiable to regard dictionaries as having authority of this kind.

To illustrate the point about commonality between dictionaries, look up the meanings of the following words in two or more modern (recently published) dictionaries from different publishers:

[1] cutaneous   javelin   polytheism   scrutinise   vegan

The *Collins English Dictionary (CED)* and the *Longman Concise English Dictionary (LCED)* have the following definitions for the words in [1]:

cutaneous	CED	'of, relating to, or affecting the skin'
	LCED	'of or affecting the skin'

javelin	CED	'1. a long pointed spear thrown as a weapon or in competitive field events. 2. the **javelin,** the event or sport of throwing the javelin'
	LCED	'a light spear thrown as a weapon or in an athletic field event; also the sport of throwing the javelin'
polytheism	CED	'the worship of or belief in more than one god'
	LCED	'belief in or worship of 2 or more gods'
scrutinise	CED	'to examine carefully or in minute detail'
	LCED	'to examine painstakingly'
vegan	CED	'a person who practises strict vegetarianism'
	LCED	'a strict vegetarian who avoids food or other products derived from animals'

In general *CED* is more wordy than *LCED*: it is a slightly larger dictionary. But you will notice how often the same words occur in the definitions in the two dictionaries. To some extent this arises from the nature of definitions (discussed in Chapter 9), but it probably also arises from the common tradition of lexicography, of which writing definitions is one aspect.

Organisation and structure

When we talk about the organisation and structure of dictionaries, much of which we will be expanding on in later chapters, we shall assume a medium-size desk-dictionary aimed at the native speaker.

A dictionary of this kind has three parts. The main body of a dictionary – and what we conceive dictionaries to be about – is the alphabetical list of **headwords** with their accompanying articles: the entries of the dictionary. The main part of the dictionary is usually preceded by what we may call **front-matter**, which the dictionary user is expected to have read before consulting the dictionary. The main part of the dictionary may be followed by a number of appendices, containing information that the dictionary editors consider may be of use to the intended users of the dictionary, though the information may not itself be strictly 'lexical'.

Now look at the front-matter of your dictionary and make a list of the material contained in it, e.g. Preface (what does it say?), Pronunciation Key, etc.

The front-matter is likely to include a list of the editorial staff responsible for the dictionary, as well as a list of contributors or specialists consulted on particular general varieties of English (e.g. American English, Australian English) or on technical varieties (e.g. astronomy, aeronautics). Such a list is included to show that the editors have engaged the help of outside specialist consultants, because they can provide the detailed specialist knowledge that in-house editorial staff cannot be expected to be familiar with. There may be a foreword or preface, perhaps explaining how the dictionary differs from the previous edition or from other similar dictionaries. Then, and perhaps most importantly, the front matter will contain instructions on how to use the dictionary. This may take the form of either annotated sample pages of the dictionary or notes and discussion of a more explanatory kind, or indeed both kinds of instruction. Dictionaries vary in the amount of instruction that they give their users: some either assume general familiarity with dictionaries on the part of users or have kept the apparatus of the dictionary deliberately simple; others, often those which have introduced innovations or have a relatively complex apparatus, have quite extensive usage notes. For example, *Chambers Twentieth Century Dictionary* has just one-and-a-half pages of 'Notes to the User', while the *Longman Concise English Dictionary* has two-and-a-half pages of explanatory charts and ten-and-a-half pages on 'How to use this Dictionary'.

Keys to enable the user to interpret the pronunciation symbols and the abbreviations used in the dictionary may also be included in the front-matter, though sometimes these form appendices or are found on the inside front or back cover of the dictionary. One further item may be found in the front-matter of dictionaries: some of the larger desk-dictionaries may contain essays on aspects of the English language. *Webster's Collegiate Dictionary*, for example, has an essay on the history of English. And the *Collins English Dictionary* has essays on 'The Pronunciation of British English', 'The Development of English as a World Language' and on 'Meaning and Grammar'. Probably most of the front-matter in a dictionary remains unread by most dictionary users, and yet much of it is intended by the editors to make the use of the dictionary easier and richer.

The same is probably also true of the appendices in a dictionary: many dictionary owners do not know what extra information is available to them in their dictionary. Look now at the appendices in your dictionary and make a list of the material contained in them.

An appendix that many dictionaries carry is one containing abbreviations which are established in the language. Many abbreviations are perhaps used regularly only by certain groups of speakers, and the same abbreviation may stand for different items for different groups; e.g. *adj.* stands for *adjective* for linguists, but for *adjustment* for bankers, and for *adjutant* in the armed forces. Another appendix commonly included in dictionaries contains 'foreign words and phrases', i.e. words and expressions – mostly from French and Latin – that are used in their original form in English speech and writing; such as 'cela va sans dire' (French, 'that goes without saying'), 'quod erat demonstrandum' (Latin, 'which was to be proved') – sometimes abbreviated to *QED* and used in proofs of mathematical theorems. Rather than carry abbreviations and foreign words and phrases in appendices, some dictionaries (e.g. *Collins English Dictionary*) include these in the main body of the dictionary in the appropriate place in the alphabetical list.

After these two kinds of appendix, which we might regard as being more strictly lexical and thus an extension of the dictionary entries themselves, dictionaries vary in the information that they include as appendix material. Tables of weights and measures occur – on the inside back cover of *Collins English Dictionary* (the only thing approximating to an appendix in that dictionary). So do lists of personal

names, names of the counties of Great Britain or the states
of the United States, the books of the Bible or the works
of Shakespeare, the monarchs and prime ministers of Britain
or the presidents of America, and so on. *Chambers Twentieth
Century Dictionary*, for example, has twelve appendices
covering some fifty-eight pages, including – in addition to
some of those already mentioned – musical terms, signs and
abbreviations; the Greek alphabet; the Russian alphabet;
Roman numerals; mathematical symbols; and a supplement
of new words coined since the previous printing of the
dictionary.

Let us turn our attention now to the main body of the
dictionary. As we mentioned earlier, this consists of an
alphabetical list of the headwords and their accompanying
articles. Although an alphabetical listing may not always be
the most appropriate arrangement of items in a lexical
description (see Chapter 14), since it gives the impression
that the vocabulary of a language consists of a set of semantically isolated words, it does have the advantage of ease of
reference, if you have a particular item that you wish to
look up. We are used to reference works of all kinds being
organised alphabetically, from telephone directories to
encyclopaedias.

We have referred to the items, usually printed in bold
type, which initiate the entries in a dictionary, as the 'headwords'. In the case of single-word lexemes (see Chapter 1,
p. 11) the citation form is used, i.e. the base form
(first/second person present, alias infinitive without *to*) of
the verb, the common singular form of the noun. Some
headwords may be multi-word lexemes, such as phrasal or
prepositional verbs (if these are treated as separate headwords) or fixed expressions (*hammer and tongs*), or
compounds written as separate words (*household cavalry*). In
some dictionaries not all the headwords may be 'words' or
even lexemes: some dictionaries include prefixes (*re-, un-*),
suffixes (*-ment, -able*), and combining forms for neo-classical
compounds (*geo-, tele-, -phile*) in the alphabetical list of
'headwords'. Also included among the headwords may be
abbreviations (*fin.* for 'financial', 'finish'), and proper names
of places (*Finland, Finisterre*) and people (*McCartney, Paul*),
as in *Collins English Dictionary*. Proper names we might
regard as being more encyclopaedic than lexical in nature,

though the dividing line between these two kinds of information must be regarded as fuzzy.

Spelling and meaning

We turn now to look at the entries themselves and the kinds of information that are provided for the headwords of the dictionary. Two kinds of information for which a dictionary is frequently consulted are to check the spelling of a word and to find out the 'meaning' of a word (see Chapter 13, p. 194). There is no tradition of spelling dictionaries in English, whereas in German-speaking countries the most widely bought dictionary is the *Duden Rechtschreibung*, which is essentially an orthographical dictionary: common words of everyday use are not provided with definitions. English-speakers use a general-purpose dictionary to check spellings. Indeed the alphabetical ordering of the dictionary is based on spelling (rather than on pronunciation); so spelling information is given willy-nilly in a dictionary. Checking a spelling may, however, not be as straightforward as one might expect; we will discuss this further in Chapter 13.

Since the organisation of a dictionary is based on the spelling of words, a first requirement in the description of the meaning of words is to distinguish homonyms and homographs, i.e. lexemes which have the same spelling. Dictionaries usually differentiate homographs by means of a superscript numeral either before or after the word concerned. For example, three homographs of *limp* are listed in the *Longman Concise English Dictionary* as *1limp*, *2limp*, *3limp*. *1limp* is the verb meaning 'to walk as if with an injured leg'; *2limp* is the noun referring to 'a limping movement or gait'; and *3limp* is the adjective meaning 'lacking firmness, not rigid'. It will be noted that the differentiation of the homographs of *limp* corresponds to their different word-class membership: verb, noun, adjective. Now examine your dictionary to discover how many homographs of *limp* are distinguished, i.e. as separate headwords.

CED, for example, distinguishes two homographs of *limp*: *limp1* includes both the verb and noun meanings of *LCED* and *limp2* is equivalent to *3limp* in *LCED*. Clearly, then, dictionaries do not necessarily agree with each other in recognising the same number of homographs for a word-form. In fact, the disagreement is between assigning different 'meanings' to homographs on the one hand, or to polysemy on the other. *LCED* is saying that the word-form *limp* corresponds to three different lexemes and must therefore be represented by three different headwords in the dictionary. *CED* is saying that the word-form *limp* corresponds to two different lexemes and that the verb and noun 'meanings' are different 'senses' of the same lexeme; they are a case of polysemy rather than homography.

A further, and, as we have seen, related requirement in the description of the meaning of words is to deal with polysemy, i.e. to distinguish the different **senses** of a lexeme, where this is appropriate. The different senses are then dealt with under the same headword and usually numbered. For example, *LCED* distinguishes two senses of the lexeme *limelight*: sense 1 refers to the light produced by stage-lighting, and sense 2 refers to the metaphorical sense, i.e. the centre of public attention. Senses of a word may be further subdivided and are then usually marked by lower-case letters of the alphabet: we may refer to these as **subsenses**. For example, *LCED* distinguishes three senses of *liberty*: sense 1 has four subsenses, and sense 3 has two subsenses. The senses and subsenses distinguished for *liberty* are as follows:

1 a 'the power to do as one pleases'
  b 'freedom from physical restraint or dictatorial control'
  c 'the enjoyment of various rights and privileges'
  d 'the power of choice'
2 'a right or immunity awarded or granted; a privilege'
3 a 'a breach of etiquette or propriety'
  b 'a risk, chance'

Now look up the word *loan* in your dictionary. How many senses and subsenses are distinguished? Compare this with the entry in another similar sized dictionary.

Let us begin with *CED*. It distinguishes six senses for *loan*. Senses 2 and 5 each have two subsenses. The first five senses relate to *loan* as a noun, the sixth to *loan* as a verb. Sense 5 relates to the phrase *on loan*, and we should note that dictionaries often treat fixed expressions in which the headword is the central item as senses of the headword lexeme. Now let us turn to *LCED*. (Here *loan* is differentiated into two homographs: *1loan* is the noun and *2loan* the verb. *1loan* has two senses, and the first has two subsenses. Senses 1a and 1b approximate to sense 2a in *CED*, and sense 2 approximates to sense 1 of *CED*. Senses 3 and 4 of *CED*, concerned with the borrowing of words from one language to another, have no mention in *LCED*. In the case of polysemy, then, there is even greater room for disagreement between dictionaries. This may be seen even more clearly if the entries are compared for lexemes which have large numbers of senses, e.g. *long*, *lose*, *low*.

Each sense or subsense of a lexeme is provided with a **definition** (see Chapter 9), which is a description of its 'meaning', a guide to the dictionary user on the place of the lexeme in the vocabulary of English and on how the lexeme is used, in this sense, in English speech and writing. Definitions are, more accurately, lexicographers' attempts, on the basis of their observations and examination of the uses of words in context, together with what they glean from other dictionaries – sometimes, regrettably, more of the latter than of the former – to depict the 'meaning' of a lexeme or of the sense of a lexeme. What is involved in the term 'meaning'. we shall be investigating more closely in the next three chapters.

In many respects the definitions are the central part of a dictionary entry. For many people it is what they go to a dictionary for: to have a given word defined. Such an action implies a view of dictionaries as authorities on the meaning of words. Most modern lexicographers would probably claim that they were merely doing their best to provide an accurate description of the meaning or meanings of lexemes based on the best information they have available on the usage of lexemes by speakers of the language. It is probably in the writing of definitions that we can speak most directly of the craft of the lexicographer (see Chapter 15). We emphasised earlier the commonality between dictionaries; we should at this point emphasise the variation in the attempts to provide sensible definitions of lexemes. Compare the following definitions of *feminism*:

Collins English Dictionary: 'a doctrine or movement that advocates equal rights for women'

Longman Concise English Dictionary: 'the advocacy or furtherance of women's rights, interests, and equality with men in political, economic, and social spheres'

Oxford Illustrated Dictionary: 'advocacy of extended recognition of claims and achievements of women; advocacy of women's rights'

One further point about meaning in the context of the overall organisation of the dictionary needs to be made at this stage. It concerns the treatment of derived words, i.e. derivations by means of prefixation and suffixation. In the case of lexemes derived by prefixation, these are to be found as separate headwords at the appropriate place in the alphabetical listing. For example, *befriend* is not found under *friend* but as a separate entry between *beforehand* and *befuddle* (*LCED*). In the case of lexemes derived by suffixation, these are sometimes found under the headword from which they are derived as what are called **run-ons**. As we have noted, derived words are treated as run-ons in *LCED* if they are not provided with a definition, because their meaning is deducible from that of the headword. So, *friendless* is a run-on under *friend*, while *friendly* is a separate headword with its run-on *friendliness*, and *friendship* is also a separate headword. In some dictionaries (e.g. *Cassell's English Dictionary*) all the lexemes derived by suffixation are run-ons under *friend*. *Collins English Dictionary* is more akin to *LCED* in its principles; *friendship*, however, is a run-on under *friend* (without definition) instead of a separate headword. Derived words are thus sometimes seen as requiring separate definition, sometimes not. This is as true of lexemes derived by prefixation as of those derived by suffixation: so, while *befriend* is defined (in *LCED*), *unfriendly*, although a separate headword, is not provided with a definition.

Ancillary information

Definitions may be the central and perhaps most important

part of the description of a lexeme in a dictionary entry, but lexicographers traditionally include much more than just definitional information in their descriptions. We have already made extensive reference to some of this information (e.g. etymology). Look up the entry for the verb *see* in your dictionary, read through it carefully, and note down all the information that is not definitional.

We shall be discussing some of this ancillary information in more detail later on (e.g. in Chapter 10), so here we want just to survey the kinds of information given and make some comments. Let us use as our starting point the entry for the verb *see* as it is found in *CED*.

Directly after the headword *see* is contained in rounded brackets an indication of the pronunciation of the headword, in this case '(si:)'. Now, while it is unlikely that any native speakers of English would need to look up the pronunciation of *see*, there are many words which may have been encountered only in writing, whose pronunciation may not be known or may need checking. For example, just above *see* in *CED* is the word *sedum* (a kind of rock plant). Apart from knowledgeable gardeners, the pronunciation of this word is probably a mystery to most native speakers. Is the stress on the first or second syllable? Is the first vowel pronounced [se:] or [si:]? *CED* gives the pronunciation as '('si:dəm)'.

The notation used to indicate pronunciation in *CED* is the International Phonetic Alphabet (IPA), the transcription system current in modern linguistics and phonetics (see Knowles, 1987). Not all dictionaries use the IPA. Many dictionaries devise their own system, which is either an adaptation of English spelling or a system that relies on the symbols of the Roman alphabet. It is sometimes argued that, because the IPA introduces a number of new and unfamiliar symbols, a barrier is created to its use by the ordinary dictionary user. The editors of the *Longman Concise English Dictionary* have accepted this argument and their pronunciation entries 'are based almost entirely on English spelling' (p. xx): the exception is the use of the schwa vowel symbol /ə/ in unstressed syllables, such as the second syllable of *sedum* '/'seedəm/'. This no doubt reduces the number of new symbols that a dictionary user must learn and makes for a more immediately usable system, but it is at the cost of some rather cumbersome symbols, e.g. /ooh/ for the vowel of *boot*, /uh/ for the vowel of *bird*, /dh/ for the initial consonant of *they*.

The other question that arises in connection with the indication of pronunciation in dictionaries is: whose pronunciation is represented? Look up in your dictionary the pronunciation of the words:

[2] garage   geography   ghastly   giraffe   glacier   grass
    gum

Does your dictionary indicate your pronunciation of these words? If not, what differences are there?

CED claims to be representing pronunciations 'that are common in educated British English speech' (p. x). *LCED* pronunciations represent a 'standard' or 'neutral British English' accent: the type of speech characteristic of those people often described as having 'no accent' . . . an accent that betrays nothing of the region to which the speaker belongs' (p. xx). That this accent corresponds most closely to the general speech forms of southern Britain means that Midlanders and Northerners, for example, are disadvantaged. If you come from either of these regions of England, not to mention from Scotland or Wales, you may find that some of the words in [2] have a pronunciation represented that differs from your own; and even if you are a Southerner you may find that your pronunciation does not correspond exactly to that represented. Not all dictionaries indicate the /ˈgærɪdʒ/ pronunciation of *garage*. You may pronounce *geography* without the /ɪ/ after the initial /dʒ/, i.e. as /dʒɒgrəfɪ/, but that variant is unlikely to be in the dictionary. *Ghastly* is almost certainly indicated with the long /ɑː/ vowel rather than the Midland/Northern /æ/, as will the vowel of *grass*. Many Southerners pronounce *giraffe* with the short /æ/ vowel, and this variant may be indicated (as it is in *CED* and *LCED*). *Glacier* likewise has variant pronunciations of the first vowel, /eɪ/ or /æ/, and these may be indicated (as again in *CED* and *LCED*). But *gum* will almost certainly have the southern /ʌ/ vowel, with no mention of the northern /ʊ/ variant. As in so many other respects, dictionaries have a problem of space here; it is impossible 'to include all the regional and social variants' (*LCED*, p. xx), and so one 'accentless', educated accent is chosen to represent them all.

After the pronunciation, the entry for *see* in *CED* has the abbreviation 'vb.' for 'verb', indicating the word-class or part-of-speech to which the lexeme belongs. This is information of a grammatical kind, which we shall discuss in detail in Chapter 10; it is information that is traditionally provided in English monolingual dictionaries, though it is doubtful whether most dictionary owners have much use for it. More useful perhaps is the other item of traditional grammatical information: inflections. In the *CED* entry for *see* we are told that *see* has the inflectional forms *sees, seeing, saw, seen*. Two of these are irregular inflections in English: past tense *saw* and past participle *seen*. Dictionaries usually indicate irregular inflections. *CED* indicates all the inflectional forms for a headword that has any irregular inflections; *LCED* indicates only the irregular inflectional forms, so *saw* and *seen* for *see*, together with their pronunciations.

One further piece of grammatical information is given in the *CED* entry for *see*. For sense 2 it says '(when tr., may take a clause as object)', and similarly for senses 4, 5, 6, 7. This is syntactic information, indicating the kind of structures in which *see* with these senses may be found. The abbreviation 'tr.' means 'transitive', i.e. being accompanied by an object in sentence structure. Dictionaries traditionally indicate whether verbs may be used transitively or intransitively (without an object), though the kind of syntactic detail given in *CED* is unusual. This aspect of grammar will be discussed further in Chapter 10.

Entries in most dictionaries of the size of *CED* contain illustrative examples, which may either be concocted for the purpose or be quotations from acknowledged sources. For example, sense 5 of *see* in *CED*, defined as 'to ascertain or find out (a fact); learn', has the example 'see who is at the door'. Not all lexemes or all senses of a lexeme are provided with an example. The examples are intended to illustrate the use of the sense of a lexeme and thus provide support for the definition. They are usually distinguished from definitions by being in italics. Editors presumably consider that some definitions are in need of more support than others, though the criteria of selection are not always obvious.

A further item of ancillary information that *CED* contains for a few of the senses of *see* relates to restrictions on usage. These may be of several kinds. Sense 16 of *see* in *CED* reads: '(in gambling, esp. in poker) to match (another player's bet). . .'. This sense of *see* is said to be restricted to a particular field or domain of language; it has what we might call a technical meaning in the field of gambling. Sense 19, which is the idiomatic expression *see (someone) hanged/damned first* is labelled 'informal'. This sense of *see* is said to be restricted to English used in an informal context, i.e. not 'formal', but not 'colloquial' or 'slang' either. We might regard these labels as general style markers. Sense 20, which is the idiomatic expression *see (someone) right* is marked as 'Brit. informal'. Like sense 19 it belongs to informal style; additionally it is said to be typical of British English rather than of North American English. Restriction on variety may relate to regional dialect varieties as well as national varieties. We shall discuss usage further in Chapter 10.

Finally under the heading of ancillary information we must mention etymology, which we referred to extensively in Chapter 2. It is traditional in monolingual English dictionaries to indicate something about the origins and historical relations of words. *See* is an Anglo-Saxon word and its etymology is given in *CED* as: '[Old English *sēon*; related to Old Norse *sjā*, Gothic *saihwan*, Old Saxon *sehan*]'. Etymological information includes, then, both the origin of the word (Old English) and reference to cognates (i.e. words related in form) in other languages. For words borrowed into the language, the century in which the borrowing took place is given, together with the language from which the word was borrowed and any previous history of the word. For example, the etymology of *seize* is given in *CED* as: '[C13 *saisen* from Old French *saisir*, from Medieval Latin *sacīre* to position, of Germanic origin; related to Gothic *satjan* to set]'. The amount of detail contained in etymologies varies from dictionary to dictionary. Etymologies are traditionally contained in square brackets.

Exercises

1. Examine the Key to Pronunciation or discussion about pronunciation in the front-matter of your dictionary. What system of notation is used? Look up the pronunciation of the following words and work out how, according to the dictionary, they are pronounced. Is this how you would pronounce the words?

   agate   chauffeur   dimension   either   lichen   longitude paella   punctuate   strength   Uranus

2. Where are the following derived words found in your dictionary? Under the headword from which they are derived, i.e. as run-ons? Or as separate entries?

   calculator   encourage   flattish   graceless   heaviness musicologist   preeminent   rusty   survivor   vaccination

3. What usage labels does your dictionary have for the following words?

   brass (='money')   *caddy   depreciable   featly   heebie-jeebies j'ouvert   maggoty* (='annoyed')   *once-over   ritenuto   titfer*

4. Look up the word *cock* in your dictionary. How many headwords are there for *cock*? On what basis are the different lexemes differentiated? How many senses and subsenses does each headword have? Are they clearly distinguishable in meaning?

5. Look in the front-matter of your dictionary and/or examine a number of entries with multiple senses to discover what principle is used for ordering the senses.

6. Can you think of kinds of information which your dictionary might usefully include but does not?

CHAPTER 4

Words and the World

Human beings have been given the capacity to talk, to communicate with each other, to make meaningful utterances so that they are understood by other human beings. They communicate about the world in which they live, about themselves, about their thoughts and feelings, about what has happened, about what might happen or what they would like to happen, and a lot more. The primary means by which human beings communicate is language. Language organises the content of communication, what human beings want to talk about, into the sounds that are heard or the written symbols that are read. Speakers or writers 'encode' the content (or meaning) of their communication into sounds or symbols using the organising principles of grammar, while hearers or readers 'decode' the sounds/symbols in order to understand the speakers'/writers' meanings.

The study of the ways in which language 'means' is called **semantics**. There is a sense in which we cannot study any part of language – sounds, grammar, words, discourse – without being conscious that language is meaningful and that all its parts serve the purpose of communicating meaningfully. We could thus look at the semantics of grammar, the semantics of discourse, the semantics of sounds; but we are restricting our study of meaning in this book to words. In this chapter and the next three we are going to consider some aspects of word meaning, the semantics of words, beginning now with the meaning relation between the words of our language and the world of our experience.

Language and experience

The relation between words and entities that we want to talk about in our experience of the world is called **reference** or **denotation**. We say that a word 'refers to' or 'denotes' something in experience; for example, that the word *rabbit* denotes a particular kind of animal. We probably think that this meaning relation is the primary, most important, or even the only one, but, as we shall see in subsequent chapters, words enter into other meaning relations as well (see especially Chapter 5). Nor is the relation of reference a simple and straightforward one, as we shall see in this chapter.

We should note first of all that in general there appears to be no intrinsic reason why a particular word should be in a relation of reference to a particular entity. The relation between words and what they refer to is **arbitrary**. There is no obvious connection between the sound or symbol sequence constituting the word *rabbit* and the animal that is denoted by that word. After all, if we examine the words denoting this animal in other languages, we find quite different sequences of sounds, e.g. *lapin* /lapã/ in French, *Kaninchen* /kaninçən/ in German. While the relation between word and **referent** (what is referred to) is arbitrary for the great majority of words, it is not so for all. Look up the following words in your dictionary and note the etymological information.

[1] didgeridoo   grunt   plod   spit   swish   thrum

For all these words the *Longman Concise English Dictionary* (*LCED*) specifies the origin as 'imit', i.e. imitative: the sound of the word is assumed to be imitative of the sound associated with the referent, which in most of the cases is an action. Sometimes then a word is 'motivated' by the sound that its referent makes; though in the course of its history the sound of the word may change, as that word undergoes sound changes taking place in the pronunciation of words in the language generally. For example, Latin *pipire* ('to chirp') may be regarded as imitative, but from it is derived (by way of French) the English word *pigeon*, which would hardly be termed imitative any more.

It is not whole words only that may be motivated by sounds associated with the referent. We might, for example, associate a common component of meaning ('unpleasant sound made by humans') with the initial *sn-* of the words:

[2] snarl   sneer   snitch   sniff   snigger   snort   snuffle

Similarly, a meaning of 'rounded protrusion' might be associated with the *-ump* ending of:

[3] bump   clump   hump   lump   rump   tump

However, we must be careful not to assume a one-to-one correspondence between sound and meaning in these cases: there are words referring to unpleasant human sounds which do not begin with *sn-* (*squeal, belch*), just as there are words beginning with *sn-* that do not have this unpleasant meaning association (*snip, snug*). There are, though, probably more words that have an element of imitation or phonetic motivation than we usually think, and many more that were originally imitative. These 'phonaesthetic' properties of words, as they have been called, have been little investigated by linguists, though they are often well known to literary scholars interested in the way that poetry sounds (see for example Knowles, 1987, Chapter 2).

We conclude then that, with any supposed imitative origin obscured in the mists of time, the majority of present-day English words have an arbitrary relation to their referents, at least as far as their sound is concerned. A different kind of motivation may be found, for instance, with many **proper names**. Proper names, i.e. names of people, places or institutions, have a unique reference, or more accurately perhaps an intended unique reference; since if it is discovered within a particular context that a name does not refer uniquely, then additions are usually made to ensure that it does. This presumably accounts for the development of surnames in European culture: if there is more than one 'Richard' in a town or village, then some way has to be found to distinguish the referents. Proper names are often semantically motivated, in the sense that a name is often related to or derived from a 'common' word whose meaning appropriately characterises the unique referent of the name. Parents sometimes choose names for their offspring because of a supposed meaning of the name, e.g. *Thomas* = 'twin', *Jennifer* = 'fair lady'. In some cultures this is common practice: many of the Hebrew names in the Old Testament derive from 'common' words or expressions that relate to the person or place concerned; e.g. *Jacob* = 'he grasps the heel' (Genesis 25:26), *Samuel* = 'heard of God' (1 Samuel 1:20). Most place-names are similarly motivated: *Chester* derives its name from Latin *castra* = 'camp', *Stourbridge* is where the River Stour was bridged.

Proper names belong exclusively to the word–class of nouns: they denote people, places and institutions. The noun class contains far more words that do not have a unique reference: **common nouns**. Rather than referring to unique 'things', common nouns refer to classes of things. Their reference is consequently much more difficult to characterise and describe, since there is no one thing that can be pointed to in order to indicate the reference. Attempt to describe the reference of the word *window*. You may like to compare your description with a dictionary definition.

As soon as you begin to think of all the kinds of objects that we can use the word *window* to denote, it seems to become more and more difficult to characterise the reference accurately and comprehensively. As soon as we move away from unique reference this becomes a problem. What, we may ask, are the essential properties of all the objects that we use the word *window* to denote? We are confronted now not by a single referent but by a whole class of referents of different shapes and sizes and perhaps even purposes, but which must have some family resemblance for them to have the same word denoting them. Dictionary definitions of words like *window* often attempt to describe these essential properties.

Here is the definition of the first (main) sense of *window* from *LCED*:

[4] an opening, esp in the wall of a building, for admission of light and air that is usu fitted with a frame containing glass and capable of being opened and shut

We shall be examining the nature of dictionary definitions in Chapter 9. Let us for the moment consider the adequacy of this definition as a characterisation of the reference of *window*. We might object that not all windows need be capable of opening, that we may have windows in roofs, that we talk of windows in vehicles such as cars and buses, that other transparent materials than glass may be used for glazing. On the other hand, we may regard these as less central denotations of *window* or of its more accidental and less essential properties, and we should perhaps take account of the lexicographer's use of 'esp' (especially) and 'usu' (usually). Let us now compare the *LCED* definition with the one in *Collins English Dictionary* (*CED*):

[5] **1.** a light framework, made of timber, metal or plastic, that contains glass or glazed opening frames and is placed in a wall or roof to let in light or air or to see through . . .
2. an opening in the wall or roof of a building that is provided to let in light or air or to see through.

Here the notions of a window as an opening and of a window as a glazed framework are separated into two senses of the lexeme. We find in this definition some of our objections to the *LCED* definition met: roofs are mentioned, the possibility of not opening is provided for; but the application to vehicles is still ignored. However, the additional function of seeing through is mentioned in this definition and the possible materials of the frame are specified, though we may wonder if the latter is an essential property or a distinctive feature of the denotation of *window*. The fact that the definitions in [4] and [5] differ illustrates the problems associated with describing the reference of words like *window*, or indeed of any words denoting things in our everyday environment. They are of a generic nature, referring to classes of items that may differ from other members of the class in a myriad ways and yet have enough properties in common for us as native speakers of the language to use a single lexeme to denote them.

Another reason for the looseness and fuzziness of the reference relation of many words is that the vocabulary of our language in some sense reflects what we choose to name in our experience of the world, or the way in which English speakers carve up reality. We see a continuity between the windows of buildings and the windows of vehicles, and so use the same lexeme to denote both; but there is a separate word to refer to the 'window' at the front of a car: *windscreen*. At one time, of course, this was the only glazing on cars, and clearly the protective function was as important

as the seeing through function. We have retained the word
even with the application of the word *window* to the glazing
of a car, and so we now regard the front window as in some
way specialised.

It is clear that to some extent the lexemes of a language
reflect the distinctions that speakers of the language wish to
make in talking about the world they live in, and indeed
particular groups may wish to make finer distinctions than
the general run of speakers. For example, for many speakers
the lexeme *car* is sufficient to denote the several kinds of
motor vehicle that others would want to distinguish by the
terms:

[6] saloon   coupe   estate   hatchback   convertible
    roadster

Similarly, the lexeme *warship* suffices for most speakers to
refer to the various vessels that a specialist would want to
distinguish by the terms:

[7] destroyer   frigate   cruiser   aircraft carrier
    minesweeper

It is going a step further, however, to maintain, as
some have done, that the way our language carves up reality
conditions us to see the world in a particular, perhaps
biased, way (a view associated with B. L. Whorf, e.g. in
Language, Thought and Reality, 1956). No doubt there is
some truth in this assertion, but the possibility always exists
either to coin a new lexeme to denote a new insight or to
extend the reference of an existing lexeme to cover a new
insight. English speakers do not categorise snow in the same
way as do Eskimos, with their several terms to denote
different kinds of snow, but it does not mean that English
speakers cannot make the distinctions, by paraphrase
('newly fallen snow', 'snow suitable for building igloos',
etc.), if not by different lexemes. The same is true within
a single language community: specialist groups have a
vocabulary to talk about their specialism, which outsiders
have no access to, and part of the task of becoming a
specialist in any field is learning the appropriate vocabulary
or **jargon**; that is, learning to carve up that bit of reality
in a more differentiated way.

What this means in terms of the relation of reference
is that some lexemes refer generally, while others refer more

specifically. Consider the lexemes in [8] and say which have a general reference and which a more specific reference.

[8] cake-fork   cutlery   fish-knife   fork   knife   spoon
teaspoon

The word with the most general reference is *cutlery*. *Knife, fork* and *spoon* are specific kinds of cutlery, but these words in turn have a general reference in relation to the more specific *fish-knife, cake-fork* and *teaspoon*. We can arrange these words in a hierarchy of generality:

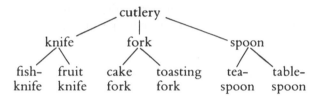

What we are illustrating here is the lexical relation of **hyponymy** (see Chapter 5, p. 64 and Chapter 14, p. 213). *Knife, fork* and *spoon* are hyponyms of *cutlery*: the meaning of *knife* etc. is included in the meaning of *cutlery*. And *teaspoon, tablespoon*, etc. are hyponyms of *spoon*. *Knife, fork* and *spoon* are, then, intermediate in the generality of their reference: they have a specific reference in relation to *cutlery*, but a general reference in relation to *fish-knife, cake-fork* and *teaspoon*.

So far we have avoided discussing the actual nature of the meaning relation of reference. This is a point of some controversy. One point of view would regard reference as a relation existing between an entity in the world and a lexeme in a language, a direct relation between 'object' and 'word'. There are two problems with this view: firstly, as we have seen, once we get away from the unique reference of proper names, the generic nature of the reference relation is less direct and more abstract; that is, the reference of a lexeme is some abstraction from the various objects in the world denoted by it. The second problem is a more acute version of the first. In our discussions so far we have used as examples **concrete** nouns only: nouns that refer to observable, tangible objects in reality. When we consider **abstract** nouns (*favour, obligation*) or verbs (*reply, overthrow*) or adjectives (*large, admirable*), it is clear that the reference relation is anything but obvious and direct.

An alternative solution is to say that the relation of reference obtains not between lexeme and real-world entity, but between a lexeme and our concept of an entity. The concept embodies all that is essential about the denotation of a lexeme. Although this has been a widely held view, it is not very satisfactory. In one way, it merely shifts the problem of explaining the reference relation from real-world object to mental concept: there are probably as many, if not more, concepts of *window* as there are objects denoted by it. Moreover, we slide from the relative objectivity of the real world to the amorphous subjectivity of the mental world. We make the **problem** abstract, not merely the referent. It would seem preferable to remain with a relatively direct reference relation, but to recognise the looseness and fuzziness in the relation that we are prepared to tolerate as users of the language.

We hinted earlier that the reference relation is not exactly the same for all word-classes. Proper nouns and 'concrete' common nouns we have discussed in some detail, since they show the relation of reference most clearly. Consider now the 'abstract' noun *purpose*, as in the sentence:

[9] My purpose in writing is to persuade you to visit us.

How would you describe the denotation of this lexeme?

We cannot point to members of a class of objects as instances of the reference of *purpose*, nor could we extract a set of essential properties in terms of shape, size or function. What we can do perhaps is to cite a number of example sentences which illustrate what a purpose is. The latter half of the sentence at [9] would be such an example of a purpose: 'to persuade you to visit us'. This makes it much more difficult to describe the denotation of such abstract nouns, and dictionary definitions often resort to finding synonyms, e.g. *LCED* for *purpose*:

[10] the object for which sthg exists or is done; the intention

where *purpose* is partly defined by the synonym 'intention'.

In the verb class the denotation of **activity** verbs, such as *blow, run, throw,* is easier to characterise than that of say **cognitive** verbs such as *believe, remember, understand.* The actions denoted by activity verbs are observable and we can abstract the essential features, as we did for generic nouns. *LCED* defines *throw,* for example, as:

[11] to propel through the air in some manner, esp by a forward motion of the hand and arm

With cognitive verbs on the other hand we are back with examples for illustration and synonyms for definition, as the *LCED* entry for *understand* shows:

[12] to grasp the meaning of; comprehend . . .

The denotation of some adjectives can be characterised quite precisely, either because they relate to words of other classes, especially nouns (e.g. *prickly* = 'having prickles'), or because they refer to observable and/or measurable qualities in the real world, e.g. colour adjectives. *Yellow,* for instance, refers to a colour of a certain hue. Other adjectives denote either relative qualities (e.g. *big, soft*) or abstract qualities (*true, furious*), and their reference cannot be described so easily. Similarly, most adverbs have an abstract reference (*bravely, lazily, fast*) and their denotation may be best characterised by paraphrase or synonym, or by relating them to words from which they are derived, usually adjectives.

When we turn from the lexical word-classes to the grammatical word-classes, the relation of reference does not apply in the same way. We noted in Chapter 1 (p. 15) that grammatical words like pronouns, determiners, prepositions and conjunctions often make no contribution to the lexical reference of a sentence. This is not always the case, however. Consider the use of the preposition *into* in [13] and [14]. What is the difference between them?

[13] We will look into your complaint.
[14] They looked over the wall into the garden.

In [13] *into* is merely a grammatical connective; it forms part of the prepositional verb *look into* (= 'investigate'), and it has no reference by itself. In [14], on the other hand, *into* refers to a particular spatial orientation; it contrasts with other prepositions denoting alternative spatial orientations (*out of, across*), but it also acts as a grammatical connective for the noun phrase *the garden*. Similarly, a conjunction like *although* may connect a subordinate clause to a main clause (grammatical function), but it also denotes a particular kind of connection ('concession' = semantic function). On the other hand, the conjunction *and* sometimes has a merely connective (grammatical) function, e.g. in:

[15] Mary made the tea and Jim read the newspaper.

This reinforces the point made in Chapter 1 (p. 17) that there is a gradation from fully lexical to fully grammatical word-classes: members of fully lexical word-classes always enter a meaning relation of reference, members of fully grammatical word-classes do not, while members of intermediate classes sometimes refer and sometimes do not, or partly refer and partly do not.

Denotation and connotation

A distinction is often made in talking about the meaning of a word between its **denotation** and its **connotation**; both of them concern the relation of a word to the world. Denotation is what we have been discussing so far in this chapter. Connotation relates to the associations that a word has over and above its denotation. Linguistically significant are the associations that a word carries for a whole language community or at least for a defined group within a language community. For example, the word *caviar* denotes 'the salted roe of large fish (e.g. sturgeon)', but it may be said to connote luxury, high living and sumptuous food. What would you state as the connotations of the following words?

[16] candle   faraway   milk   pig   tram

For many people *candle* may have religious connotations or alternatively romantic associations as lighting for an intimate meal. *Faraway*, which is denotationally synonymous with *distant*, has romantic connotations absent from the latter. *Milk* for many will connote health and strength, especially if you belong to the generation that consumed one-third of a pint every school-day; more recently perhaps the connotation has changed to the opposite, at least with the health and slimming conscious, especially in view of the association of dairy products with heart disease. *Pig* no doubt connotes uncleanness and unpleasant smells for many. *Tram* on the other hand may have connotations of nostalgia or holidays.

Can you find additional connotations of these words, apart from those just mentioned?

Two points need to be noted from our discussion of connotation so far. One is that connotations are far more indeterminate than denotations. On the one hand connotations may be subject to considerable variation from one generation to the next (e.g. *milk*, or consider what the word *siren* means to the generation that experienced the Second World War). On the other hand connotations may be rather subjective and not shared in the same way by all speakers of a language: our individual experience of language and its relation to the world is to some extent unique and idiosyncratic. The other point about connotation is the extent to which it relates to the lexeme itself rather than to the entity that the lexeme denotes, or whether it is not possible entirely to separate the two. The connotations of a word for us must reflect our experience of the entity to which the word refers and the place which this entity has in our belief-systems and thought-patterns; but we no doubt transfer the associations of the entity itself to the lexeme we use to denote it. Besides, connotations shared by a group of language users or a whole language community are part of the cultural package that we inherit with the language itself.

Another term that is associated with connotation, though rather narrower in scope, is **emotive meaning**. In any culture at any time there are words which are used by sloganisers, political or otherwise, to stand, some for positive and some for negative values, judged as such by that culture. Sometimes these 'emotive' overtones have been more important in a word's use than the denotation of the word. We may cite as examples words such as the following:

[17]   imperialism   revolution   freedom   democracy
       republic   justice   equality   progress   rights   law

and you can probably think of more to add to the list. After all, both communist and capitalist systems call themselves 'democratic'. Similarly, advertisers have a series of adjectives (and images) whose emotive meanings all but eclipse their denotations, such as:

[18] modern   delicious   special   fine   real   fresh
      pure genuine healthy

The referential meaning of a lexeme therefore, its denotation and connotation, depends on the context of its use, in two senses. First of all it depends on its linguistic context, the other words in the same sentence, paragraph or even text. The lexeme *freedom* has a different meaning in the context of *prisoner, gaol, cell, sentence* and *warder* than it has in the context of *oppression, dictator, injustice, regime* and *junta*. Secondly, the referential meaning of a lexeme depends on its situational context: who is using the word, who the audience is, what the occasion of use is. *Freedom* for a dissident in the Soviet Union might mean the capacity to read, hear, say and write what he or she thinks; in South Africa, on the other hand, freedom for a black person might mean the lifting of apartheid restrictions on movement and residence. In both contexts the 'freedom from oppression' sense is meant, but the situation determines the meaning more precisely. For some words it may be the connotation that is affected by context: in choosing clothes the word *stylish* may have a positive connotation for one person but a negative connotation for another, cf. the phrase 'too stylish'.

Clearly, the definitions of lexemes that we find in dictionaries do not, indeed cannot, take account of the kinds of variation in referential meaning that we have been discussing. The division into senses is an attempt to take account of context, but the description of connotation by and large finds no place in dictionary definitions. The meaning of (the sense of) a lexeme that we find given in a dictionary entry must therefore be regarded as 'potential', a distillation of the essentials, awaiting actualisation in a particular linguistic and situational context. Because this is so, we cannot cite a dictionary definition as authoritative in the interpretation of a word in a particular context. We must allow for the possibility that a word will be used in new contexts with new 'meanings'.

Denotation and definition

To conclude this chapter we will now bring together some observations on the relation between the denotation of lexemes and their definitions in dictionary entries. First of all, look up and examine the definitions of the following items in your dictionary:

[19] salt   decorate   astute   offhand

The definition of *salt* is likely to contain the chemical term 'sodium chloride' and perhaps even the chemical formula 'NaCl'. It is unlikely that any language user would recognise the substance in the salt-cellar from this definition: is this 'scientific' definition a true reflection of the denotation (not to mention connotation) of this lexeme? It must be said though that the definition is likely to contain some mention of the use of salt for 'seasoning and preserving' food, which comes a little nearer to everyday use. In general we have noted that 'concrete' nouns do have definitions that are descriptions of the 'things' denoted by their lexemes; some dictionaries, indeed, contain drawings or photographs for this purpose.

Similarly in the case of 'activity' verbs like *decorate*, the definition often describes the action denoted by the verb, in this case 'to apply new coverings of wallpaper or paint' (*LCED*) or 'to make more attractive by adding ornament, colour, etc.' (*CED*). When we turn to more 'abstract' lexemes, dictionary definitions tend to be less descriptive and analytical. The definition of the adjective *astute*, for example, probably relies largely on the citation of synonyms (see Chapter 5), or a synonymous paraphrase, e.g. 'shrewdly perspicacious' (*LCED*), 'having insight or acumen; perceptive; shrewd' (*CED*). The same is true for the adverb *offhand*: 'without forethought or preparation' (*LCED*), 'without preparation or warning; impromptu' (*CED*). Some of these points will be considered in more detail in Chapter 9.

Dictionaries do not normally contain lexemes with unique reference, i.e. names of people and places. Such items are considered to be more appropriately located in an encyclopaedia. But the dividing line between dictionary and encyclopaedia is fuzzy and some dictionaries do contain such items, e.g. *Collins English Dictionary, Oxford Illustrated Dictionary*. Names of (famous) people are defined by giving their dates and the reason why they are considered famous, e.g. (*CED*):

> [20] **Sobers. . . Garfield St Aubrun.** born 1936, West Indian cricketer; one of the finest all-rounders of all time.

Names of places are defined by location and other appropriate geographical detail, e.g. (*CED*):

> [21] **Birmingham. 1.** an industrial city in central England, in West Midlands: the second largest city in Great Britain. Pop.: 1013366 (1971). **2.** an industrial city in N central Alabama: rich local deposits of coal, iron ore, and other minerals. Pop.: 295686 (1973 est.).

The unusualness of these items in British dictionaries (they are quite usual in American dictionaries) makes them a feature of the *Collins English Dictionary*, which includes fourteen thousand names of people and places.

Exercises

1. Attempt a description of the denotation of the following 'concrete' nouns. Then check your attempt with a definition in one or more dictionaries.
   cup jam (contents of jamjar) *path screw* (as in screwdriver) *wine*
2. What verbs can you think of in English that refer to the activities of (a) taking people on for work (e.g. in a business), and (b) causing people to cease from work? What differences in meaning do you notice between them?
3. What adjectives do we have in English that refer to the experience of (a) good and (b) bad smells?
4. What connotations do you think the following nouns have in English?
   charity iron mole snow street

5. Which of the following nouns has its reference described 'scientifically' in your dictionary (as *salt* = 'sodium chloride')?

    badger   mushroom   robin   sodium   spider

6. Look up the following sets of words in your dictionary. Do they have synonym-type definitions? And are they at all defined circularly, in terms of each other? Can this be a disadvantage of synonym-type definitions?

    *enjoyment – gratification – pleasure   abuse (verb) – revile
    confidence – faith – trust   frighten – scare   florid – flowery
    – ornate*

CHAPTER 5

Words and Words

In the previous chapter we considered the meaning relation that holds between words and the world, extra-linguistic reality as it is sometimes called. This relation by no means exhausts what we can say about the meanings of words; indeed we saw that dictionary definitions often find it difficult to characterise the meanings of some words in referential or denotational terms. Another aspect of meaning and the study of meaning (semantics) is the meaning relations that hold within the vocabulary of a language between words themselves: **lexical relations** or, as they are often called, **'sense' relations**. The meaning of any lexeme may be described, then, both in terms of its reference or denotation and in terms of its sense relations: both contribute to characterising a lexeme's meaning. In this chapter we are going to examine two kinds of sense relation that may occur between words.

The two sense relations concerned are **synonymy** and **antonymy**. The term 'synonymy' derives from Greek, and its two parts (*syn-* + *-nymy*) mean 'same + name': synonymy deals with sameness of meaning, more than one word having the same meaning, or alternatively the same meaning being expressed by more than one word ('name'). 'Antonymy' likewise derives from Greek, and its two parts (*ant-* + *-nymy*) mean 'opposite + name': antonymy deals with oppositeness of meaning, words with opposite meanings of various kinds. As we shall see, these two sense relations, in spite of their obvious connection (same – opposite), are very different in kind.

One further sense relation should be mentioned here, though we shall discuss it later in another context in Chapter 14, p. 213: **hyponymy**. Again this term derives from Greek, and its two parts (*hyp-* + *-nymy*) mean 'under + name'. Hyponymy refers to the hierarchical relationship between the meanings of lexemes, in which the meaning of one lexeme is included in (under) the meaning of another lexeme. This is seen most obviously in scientific classifications such as we find in biology, where for example the meaning of *panther* is included within that of *cat* (as the name of the cat family). In its sense of 'domestic cat' the word *cat* is itself a hyponym of the general word *cat*.

Synonymy

Two words are said to be synonyms if they 'have the same meaning'. It is difficult to understand how a word 'has' a meaning. As we saw in Chapter 4, the description of meaning in a dictionary definition is an indication of the meaning potential of a word: only in a linguistic and situational context is the meaning actualised. Synonymy therefore needs to be defined in terms of contexts of use: two words are synonyms if they can be used interchangeably in all sentence contexts. Consider the following pairs of 'synonyms'. Can you think of any sentence context in which one member of a pair may be used and the other member not? Make sentence frames to illustrate this point, e.g. in the frame 'I am not at _____ to tell you', the word *liberty* may be inserted but not its synonym *freedom*.

[1] discover – find   keep – retain   busy – occupied
frequently – often   decoration – ornamentation

Discover and *find* are synonymous in a sentence like 'We found/discovered the boys hiding in the shed', but *find* could not substitute for *discover* in 'Sir Alexander Fleming discovered penicillin in 1928'. *Keep* and *retain* are synonymous in the sentence 'Keep/retain your ticket for further inspection', but *retain* could not replace *keep* in 'We keep the door locked all night'. *Busy* and *occupied* are synonymous in the sentence

'I'm afraid Mr Smith is busy/occupied at the moment', but *busy* could not substitute for *occupied* in 'I'm afraid this seat is occupied'. *Frequently* and *often* are synonymous in the sentence 'Do you go to concerts frequently/often?', but *frequently* could not substitute for *often* in 'You don't often see policemen sneeze on duty'. *Decoration* and *ornamentation* are synonymous in the sentence 'These porcelain vases have very fine ornamentation/decoration', but *ornamentation* could not replace *decoration* in 'She's very expert at cake decoration'.

In none of these cases do the pairs appear to be interchangeable in all contexts, since we have found at least one context for each in which one member of the pair may occur but not the other. This suggests either that our examples were badly chosen or that we have defined synonymy in an impossible way. Indeed the definition of synonymy as 'interchangeable in all contexts' is sometimes referred to as **strict synonymy**, and many linguists doubt whether synonymy of this kind occurs at all in language. There are two arguments against strict synonymy. One is economic: having two words which are totally synonymous, and even more so if there are large numbers of such pairs, is a luxury which a language can afford to do without. The economy of a language will not tolerate, except perhaps for a short period of time, the existence of two words with exactly the same range of contexts of use; and it certainly will not tolerate a proliferation of them.

The second argument against strict synonymy is the historical counterpart to the first. It has been noted that if strict synonyms occur in the language, whether by borrowing or for some other reason, then one of two things tends to happen. One is that a differentiation of meaning takes place and one of the words begins to be used in contexts from which the other is excluded, perhaps through semantic specialisation. When, for example, *mouton* was borrowed into English from French in the medieval period, it was absolutely synonymous with *sheep*. It still exists in the vocabulary of English as *mutton*, but its meaning is specialised, referring only to the meat of the animal consumed as food, while the animal itself continues to be called by the Anglo-Saxon word *sheep*. Alternatively, as we have noted before (in Chapter 2, p. 23, and see below, p. 69), one of the words in a synonym pair may be stylistically restricted. Usually, the borrowed word is associated with more formal style. This happens with *commence*, borrowed from medieval French, as against Anglo-Saxon *begin* or *start*.

The other thing that may happen to counter strict synonymy is that one of the words will fall out of use and become obsolete, leaving the other as the sole lexeme with that meaning; or it may become highly restricted, like *kith*, found only in the expression *kith and kin*. For example the word *reward* was introduced into English from Norman French, but English already had a word with the same meaning: *meed*. In this case *meed* has fallen out of use and *reward* has completely taken its place. A similar process has happened with *foe* and *enemy*, though *foe* is still retained in some contexts, mainly of a literary nature.

When we talk about synonymy we do not generally have strict synonymy in mind. We are thinking much rather of pairs of words that can substitute for each other in a wide range of contexts but not necessarily absolutely, or that we think of as having the same general reference, such as *big/large, refuse/decline, freedom/liberty, sometimes/occasionally, beneath/below*. We might contrast this kind of synonymy with strict synonymy by calling it 'loose' synonymy, with varying degrees of looseness, no doubt. And synonymy in this sense is a meaning relation that holds between a great number of lexemes in the English vocabulary.

Look up the following pairs of synonyms in your dictionary and make a note of the origin of each lexeme:

[2] help – aid   teach – instruct   heaven – sky   first – initial   kingdom – realm   annoy – irritate

---

A major reason for the existence of so many pairs of synonyms in English is the different origins of the members of a synonym pair (compare Chapter 2). In [2] all the first members of the pairs, except for *annoy*, are Anglo-Saxon words; their origin is in Old English. *Annoy* came from French. Of the second members of the pairs, *aid* and *realm* came from French, *instruct* and *initial* from Latin and also *irritate*, while *sky* came from Old Norse through the Viking invaders. In the case of the first five pairs there is a contrast between an Anglo-Saxon word and a borrowed word; in the case of the sixth pair, both words were borrowed, one from French and one from Latin. Indeed, we can find triplets of synonyms in English, representing Anglo-Saxon, French and Latin origins, as in *kingly, royal* and *regal* respectively. Interestingly it is the French word *royal* which has become the common or neutral term.

Origins may explain how the English vocabulary came to contain so many pairs of synonyms and give some idea of the semantic adjustments that must have taken place in the course of the language's history, but they do not explain how the synonyms relate to each other and are differentiated in the language now on the principle of economy that we discussed earlier. Pairs of lexemes may refer to the same entity but be differentiated in a number of ways, e.g. by being restricted in occurrence to some particular linguistic or situational context or contexts. We will now examine some of the ways in which synonyms may be differentiated.

A pair of synonyms may persist in the vocabulary because they belong to different dialects. Different groups of speakers of the same language use different words to refer to the same entity. Because the dialects are regionally bound the synonyms persist. Indeed, for the speakers concerned, unless they are bidialectal, there is no synonymy, only from the point of view of the language as a whole. In many cases though there is at least passive bidialectalism: speakers of one dialect use only one of the pair of synonyms but readily understand the other. This is the case, for example, with many pairs of synonyms in British and American English, such as the following:

[3] lift – elevator   pavement – sidewalk   sweet – candy   biscuit – cookie   tap – faucet   boot – trunk (car)   flat – apartment   dustbin – trashcan

In some cases, however, there may be misunderstanding. For example, the North American equivalent of British *braces* is *suspenders*, but the latter refers to a quite different item of clothing in British English. The misunderstanding works the other way with British English *homely*, which in American English has the meaning 'plain, ugly': the American English equivalent of *homely* is *homy* or *homey*.

There are synonym pairs not only between national varieties of the language (British – American – Australian, etc.) but also between dialects of a national variety. For many regional dialect speakers these will represent real synonym pairs if they are speakers of the standard dialect as well, while most other speakers will be unaware that such dialect pairs exist. Look up the following regional dialect words in your dictionary to discover the standard dialect synonyms (they may not be all entered in your dictionary, though they are all in the *Collins English Dictionary*):

[4] butty   culch   diddle   heartsome   lease   mullock
    pawky   snap   stob   tum

You will notice that some of these regional dialect words have standard dialect homonyms (e.g. *lease*). The standard dialect synonyms of the words in [4] are given in [5], as found in *CED*.

[5] butty – sandwich   culch – rubbish   diddle – jerk
    up and down   heartsome – cheering   lease –
    common land   mullock – mess/muddle   pawky –
    with a dry wit   snap – packed lunch   stob – post
    tum – empty

A second way in which synonyms may be differentiated is by style or level of formality. We noted in Chapter 2 (p. 23) a general preponderance of Anglo-Saxon words in colloquial language and an increase in French- or Latin-derived words in more formal kinds of speaking and writing. In many instances there are pairs of synonyms that are differentiated by stylistic level, one member of which is an Anglo-Saxon word (the colloquial one), and the other is a word originally borrowed from French or Latin (the formal one). For example, of the synonym pair *climb – ascend*, *climb* is Anglo-Saxon in origin, while *ascend* entered English from Latin in the fourteenth century. Similarly, *go in* is Anglo-Saxon in origin, while *enter* came into English from French in the thirteenth century.

Here now is a list of Anglo-Saxon words that we might associate with colloquial language. Suggest a more formal synonym for each of them and find out the origin.

[6] begin   before   burn   funny   gift   kiss   last
    (opposite of *first*)   odd   stop   think

A more formal synonym of *begin* is *commence*, borrowed from French in the fourteenth century. *Before* has the synonyms *preceding*, from Latin via French in the fourteenth century, and *previous*, from Latin in the seventeenth century. In this example we can see a tendency that is reflected in many synonym triplets: a medieval borrowing from French complements an Anglo-Saxon word, and is in turn complemented by a Renaissance borrowing from Latin. Sometimes the same word is borrowed twice in different guises; e.g. *please* was borrowed from French in the fourteenth century (as a synonym of *make glad*), and it derives from Latin *placere* ('to please'), which was borrowed directly into English in the seventeenth century as *placate*.

The next word in [6], *burn*, has the synonym *incinerate*, borrowed from Latin in the sixteenth century. *Funny* has the synonym *amusing*, from French in the fifteenth century. *Kiss* is complemented by the more formal *embrace*, from fourteenth-century French. *Last* has two synonyms from Latin: *final*, borrowed in the fourteenth century, and *ultimate*, borrowed in the seventeenth century. *Odd* likewise has two synonyms: *strange*, from thirteenth-century French, and *peculiar*, from Latin in the fifteenth century. *Stop*, like *begin*, has a fourteenth-century French synonyms: *cease*. And *think* has three synonyms: *ponder*, from French in the fourteenth century, *consider*, from Latin in the fourteenth century, and *cogitate*, from Latin in the sixteenth century. The dates are from the *Collins English Dictionary* and refer to the first recorded instance found by lexicographers and scholars of etymology. These examples point to an extensive range of synonymous pairs of words that are differentiated by stylistic level and reflect the fact that we can, by appropriate selection of vocabulary, choose the level of formality at which we pitch a text or discourse.

A third related means by which synonyms are differentiated, but one which is usefully distinguished from style, is technicality. Many professions, trades, sports and hobbies have developed vocabularies which contain lexemes appropriate to the activity engaged in but which are not part of everyday language. We refer to such lexemes as **technical vocabulary** or **jargon**. In many cases technical words are necessary as a means of talking precisely about aspects of the activity concerned, and distinctions are made that nonspecialists have no need to make. In some cases, though, a technical word may have a common-language synonym: use of the jargon serves as a mark or badge of membership of the profession, trade, etc. and perhaps as a means of mystification for non-specialists. For example, medical specialists may refer to matters related to the *lung* by means of the Latin-derived adjective *pulmonary*, an eighteenth-century borrowing, e.g. 'pulmonary disease' = 'lung disease'. Doctors talking to patients need, therefore, to translate their technical jargon into the words of everyday language.

Look now at the list of technical words in [7] and suggest an ordinary language synonym for each of them.

> [7] cardiac convulsion cranium incision lesion
>   mamillary neurosis ocular ophthalmic optic
>   patella psychotic trachea; auditory lexeme
>   orthography phoneme semantic

The first thirteen words in [7] come from the jargon of the medical profession, which is noted for its proliferation of technical vocabulary based on Latin and to a lesser extent on Greek. Latin, after all, was for a long time the language of communication for medical scientists, and biological classifications were originally undertaken in that language. It also serves as a suitable source of mystification in the doctor–patient relationship, though a health- and fitness-obsessed generation has taken many of these terms into the common vocabulary. The last five items are a selection from the perhaps familiar jargon of linguistics.

In detail, the ordinary language synonyms of the items in [7] are as follows: cardiac – heart, convulsion – fit, cranium – skull, incision – cut, lesion – injury, mamillary – breast, neurosis – anxiety, ocular/ophthalmic/optic – eye, patella – kneecap, psychotic – mad, trachea – windpipe; auditory – hearing, lexeme – word, orthography – spelling, phoneme – sound, semantic – meaning. It is interesting to note the three technical synonyms of *eye*. All three were borrowed into English in the sixteenth century from Latin. Two of them, though, originate from Greek, which had two words for *eye*: *optilos*, which gave us *optic*, and *ophthalmos*, which gave us *ophthalmic*. The Latin word for *eye, oculis*, gave us *ocular*. Within medical jargon the three terms would appear not to be used interchangeably; e.g. the nerve in the eye is called the *optic nerve* and not the *ophthalmic* or *ocular nerve*. And an 'ophthalmologist' is not the same as an 'optician': the former diagnoses and tests eye diseases, while the latter supplies spectacles. However, opticians who undertake both kinds of work may call themselves 'ophthalmic opticians', or 'optometrists'. It is doubtful whether the general public understands the distinctions that are being made here.

A fourth way by which synonyms may be differentiated is connotation. One member of a pair of synonyms may have connotations not shared by the other member. For example, *love* and *adore* could be said to be synonyms, but *adore* has connotations of passion or worship, which *love* does not share: *love* is the more neutral of the pair. Similarly, *devise* and *contrive* are synonyms in one of their senses, e.g. in 'He devised/contrived a solution to the problem', but *contrive* has the connotations of subtlety or even trickery, while *devise* is more neutral.

Now consider the groups of synonyms in [8] and say how the members of each group differ in their connotations.

[8] crowd – mob   pleased – delighted   look at – stare at – gaze at   modern – up-to-date   boring – monotonous – tedious

---

In each case the most neutral lexeme has been placed first, the others being synonyms with special connotations. *Mob* connotes a crowd with riotous tendencies and is pejorative. *Delighted* has connotations of being extremely or intensely pleased. *Stare at* connotes a looking which is intent and possibly hostile; *gaze at* has more the connotation of wonderment or admiration. *Up-to-date* connotes a modernity that is fashionable. *Monotonous* has a pejorative connotation in relation to being boring, and *tedious* is even more pejorative in its connotation. Our attitude in respect of what we are talking about may be reflected in our vocabulary choice among synonyms that have different connotations.

A fifth reason for the existence of synonyms, or means by which they may be differentiated, is **euphemism**. There is a taboo, in some contexts at least, on referring directly to certain subjects, especially death, sex and some bodily functions. Consequently, euphemistic synonyms have been coined to refer more obliquely to these taboo subjects. Interestingly, these topics usually have colloquial or slang synonyms too. For example, the euphemistic synonym of *die* is *pass away*; and the colloquial synonym (or **dysphemism**) is *snuff it* or *kick the bucket*, which might be termed idioms (see Chapter 7, p. 106). Similarly, *kill* has the euphemistic synonym *liquidate* and the colloquial synonym *do in*.

Another human condition, self-inflicted this time though, which attracts euphemistic synonyms is drunkenness. A euphemism, though actually perhaps a more formal term, for *drunk* is *intoxicated* or *inebriated*; and there are numerous slang or colloquial synonyms, e.g. *sloshed, sozzled, stoned, pissed. Urinate* has the euphemistic synonym *pass water*, and the dysphemism *piss*, or, felt to be less vulgar or offensive, *pee* and with young children *wee*. The *buttocks* are referred to euphemistically as the *behind* and colloquially as the *bum*, or more vulgarly *arse*. These examples will suffice to illustrate the point. Clearly synonym pairs arising from euphemism are restricted in number and confined to limited areas of human experience.

The synonyms that we have discussed so far have been pairs or triplets of lexemes that have more or less the same reference but which differ in their contexts of use: geographically (dialect), stylistically (informal vs. formal), in domain or register (technical vs. common), attitudinally (connotation), or in sensitivity (euphemism). There is a further kind of synonymy that we might recognise, which is sometimes called 'partial' synonymy. Part of the meanings of two (or more) words are the same: there is overlap in meaning but not complete identity of meaning. For example, the meanings of the lexemes *mature, ripe, adult* overlap: they all refer to growth having been achieved to a certain point (of maturity), but they each refer to something more than that. *Ripe* infers 'ready to eat' as well as 'mature'; *adult* infers 'responsible, no longer a child' as well as 'physically mature'; *mature* has the most general reference of the three, but infers 'wise, sound in judgement' in relation to human beings and 'stored long and well' in relation to wine.

Group the words in [9] into triplets of lexemes with overlapping meanings, i.e. sets of partial synonyms.

[9] brim   crush   decorate   edge   enlist   genuine
    hire   income   make up (verb)   mash   paint
    pound (verb)   real   recruit (verb)   rim   salary
    sincere   wages

The lexemes in [9] may be grouped into the following sets of partial synonyms: brim – edge – rim, genuine – real – sincere, enlist – hire – recruit, decorate – paint – make up, crush – mash – pound, income – salary – wages. Once we begin to take account of synonymy and overlapping meanings, it is clear that the sense relation of synonymy is a very important principle in defining the meanings of large parts of the vocabulary of English, and as we shall see (later in this chapter and in Chapter 9, p. 135), it is used extensively in the definition of lexemes in the dictionary. It is also an important principle in establishing and describing lexical fields (see Chapter 14).

Antonymy

We turn now to antonymy, a sense relation of a quite different kind than synonymy. Antonyms are not differentiated for formality or dialect or technicality: antonyms occur within the same style, dialect or register. But the relation of antonymy is not uniform; there are different kinds of antonym. As a beginning to our discussion, list the antonyms of the following lexemes:

[10] alive   male   narrow   open   over   receive
     relinquish   sell   small   tall   weak   wife

It will be noted that a number of word-classes are represented by the lexemes in [10]: verb (e.g. *receive, sell*), noun (*wife*), preposition (*over*), adjective (e.g. *alive, narrow*). There is a preponderance of adjectives, however; and it is in this word-class that the relation of antonymy operates most widely. We might list the antonyms of the items in [10] as follows: alive – dead, male – female, narrow – wide, open – shut, over – under, receive – give, relinquish – retain/keep, sell – buy, small – large/big, tall – short, weak – strong, wife – husband.

A careful examination will reveal three kinds of oppositeness of meaning represented by the pairs of antonyms discussed. We can group them into three sets, with four pairs of antonyms in each set, as follows:

[11] narrow – wide   small – large   tall – short   weak – strong

[12] alive – dead   male – female   open – shut   relinquish – retain

[13] over – under   receive – give   sell – buy   wife – husband

The antonyms represented in [11] are called **gradable** antonyms. They are adjectives which do not refer to absolute qualities, but which may be subject to comparison or qualification. For example, we could say of a road that it is 'very narrow' or 'very wide', 'quite narrow' or 'quite wide', or that one road is 'wider' or 'narrower' than another. Moreover, the reference of the adjective is relative to the noun that it is modifying; e.g. the width or narrowness of roads is within a different set of parameters than the width or narrowness of, say, footpaths or ribbons. Another interesting feature of gradable antonym pairs is that if you wish to ask questions about the quality concerned only one of them is normally used. In size adjectives it is the 'large' one; e.g. we normally ask 'How wide is the road?' not 'How narrow is the road?'. Indeed to ask the latter implies a prior identification of narrowness and that the narrowness is somehow crucial to what is being talked about. Similarly in answering the question 'How wide?' we say 'Three metres wide' not 'Three metres narrow'.

The antonyms represented by the pairs in [12] are called **complementary** antonyms. Complementarity means that the denial of one member of the pair implies the assertion of the other member. If *not X* implies *Y*. If someone is *not dead* they are *alive*; if a person is *not male* then they are *female*; if the shop is *not open* it is *shut*; if you do *not relinquish* a post you *retain* it. There is a more clear-cut either/or opposition with complementary antonyms than with gradable antonyms, though the distinction between the two types is perhaps rather more fuzzy than we have implied. Someone can after all be 'more dead than alive', or even 'very much alive'; and a door may be 'almost shut' or 'not quite open'. Indeed, just about any non-gradable antonym may be made gradable. The at one time seemingly clear-cut *male/female* distinction is now called into question by sex-change operations and advances in knowledge about chromosomes. Conversely, if a road is *wide* it is *not narrow*, though we might discuss whether we would call the road in question 'wide' or 'narrow', and we could argue about how wide or narrow it is. Nevertheless there is a distinction worth making here.

The antonyms represented by the pairs in [13] are called **converses** or **relational opposites**. One member of the pair refers to the converse relation referred to by the other member. For example, if the bathroom is *over* the hall, then the hall is *under* the bathroom. Similarly, if Mary *receives* chocolates from Bill, then Bill *gives* chocolates to Mary; if Harry *sells* chocolates to Bill, then Bill *buys* chocolates from Harry; and if Mary is Bill's *wife*, then Bill is Mary's *husband*. A relation exists between the antonyms such that one is the converse of the other: they represent two (opposite) perspectives on the same relation. This type of antonymy is quite distinct from the other two and there appears to be no overlap.

Sense relations and definitions

We stated at the beginning of this chapter that sense relations have as important a contribution to make to the meaning of a lexeme as the relation of reference. In that case it is reasonable to expect lexicographers to give attention to and utilise sense relations in their definitions of lexemes in the dictionary. Indeed we noted in Chapter 4, (p. 61) that for many lexemes the denotation is not easily or not at all possible to express in other words: the lexicographer is perforce obliged to use sense relations or some form of paraphrase for definitions. Above all, the relation of synonymy is extensively used in dictionary definitions, antonymy to a lesser extent.

Look up the definitions of the following words in your

dictionary. To what extent are synonymy and antonymy used in the definitions?

[14] chancy   deceased   elapse   fed-up   greed
     main   (adjective)   ordinary   unkempt   wanting

As these lexemes are defined in the *Longman Concise English Dictionary* we find both synonymy and antonymy being used in the definitions. *Chancy* as 'uncertain . . . risky' (synonyms), *deceased* as 'no longer living' (antonym), *elapse* as 'to pass by' (synonym), *fed up* as 'discontented, bored' (synonyms), *greed* as 'excessive acquisitiveness; avarice' (synonyms), *main* as 'chief, principal' (synonyms), *ordinary* as 'routine, usual' (synonyms) and 'not exceptional' (antonym), *unkempt* as 'not combed; dishevelled' and 'not neat or tidy' (antonyms and a synonym), and *wanting* as 'not present' (antonym). We could go so far as to say that the lexicographer could not do without involving the sense relations of synonymy and antonymy, as well as that of hyponymy, in definitions, in particular synonymy. We shall have occasion to return to the question of sense relations in definitions in Chapter 9.

Exercises

1. Consider the following pairs of words. Why are they no longer synonyms?
   *chamber – room   fleer – grin/laugh   reck – mind/care
   sooth – truth   spirit – ghost   doom – judgement*
2. How is each member of the following pairs of synonyms differentiated from the other?
   *bale – bundle   cicatrix – scar   depression – slump   gowk
   – fool   lumber – logs   naturism – nudism   remuneration
   – pay   sufficient – enough   teem – abound   umbilicus –
   navel*
3. What kind of antonymy is represented by each of the following pairs of antonyms?
   behind – in front   captive – free   fast – slow   fixed – loose

   high – low   in – out   leave – stay   north of – south of
   parent – child   rich – poor   teacher – pupil   thin – fat
4. Which of the following lexemes are defined in your
   dictionary using the semantic relation of synonymy?
   back out   banjo   barely   barring   beaker   beastly
   biff   bitty   blancmange   bland

CHAPTER 6

Analysing Word Meanings

In the previous two chapters we have considered two kinds of meaning relation that words enter into. One, the referential or denotational relation, concerns the relation between words and our experience of the world, what we want to talk about. The other, sense relations and in particular the relations of synonymy and antonymy, concerns the relations that words contract with each other in the meaning systems of the vocabulary of the language. Neither kind of relation by itself exhausts what we can say about the meaning of a word, and the meaning relations are not entirely independent of each other either. Synonymy after all is about sameness of reference, and discussion of the reference of a lexeme cannot be undertaken in isolation from the consideration of semantically related lexemes.

What we have not yet attempted is an explanation of meaning relations by attempting an analysis of the meanings of lexemes. Such an analysis might look at what aspects of meaning are common to a group of semantically related lexemes, and what aspects serve to make distinctions of meaning among the lexemes. We might also ask whether these aspects of meaning could be generalised to other groups of lexemes, and whether such an approach could be applied to all lexemes in the vocabulary of a language. Analysis of this kind has been undertaken under the heading of **componential analysis**: the meanings of lexemes are analysed into components, which can then be compared across lexemes or groups of lexemes.

Componential analysis

The idea of dividing the meaning of a lexeme up into a number of components of meaning has a parallel in phonology, where a speech sound (phoneme) is described in terms of its **distinctive features** of sound. For example, the sound /b/, as in 'ball', is described as 'voiced' (with vocal cords vibrating), 'bilabial' (made with the two lips) and 'stop' (involving a complete constriction in air flow). It shares each of these features (or components) with other sounds in the phoneme inventory of English, but together they uniquely characterise /b/. Hence, the features are 'distinctive' in that they serve to distinguish one from all the other sounds of English. They do not necessarily say all that there is to be said about the sound of a phoneme, but they do serve to distinguish one phoneme from all the others in the inventory. Now, by comparison with the lexeme inventory, the phoneme inventory (some forty sounds) is extremely small, and the phonemes of English can be distinguished by a quite small number of features.

It is argued that some of the vocabulary of English (or any language) can be similarly analysed to produce a set of components which will distinguish the meanings of all the lexemes in the language. For example, the meaning of the lexeme *girl* is said to be distinguished by the components 'human', 'non-adult', 'non-male'. The component 'human' distinguishes *girl* (and lexemes such as *boy, woman, man*) from non-human creatures; 'non-adult' distinguishes *girl* (and *boy*) from *woman* and *man*; 'non male' distinguishes *girl* (and *woman*) from *boy* and *man*. Components such as these are sometimes presented as either/or (i.e. **binary**) choices, conventionally written in capital letters and placed in square brackets, e.g. [+ADULT]/[−ADULT]. The meaning of *girl* would then be expressed by the components [+HUMAN], [−ADULT], [−MALE]. It will be noted that such components represent complementary antonyms. We will find, however, that components may not always be expressed in terms of binary choices.

Using the components we have discussed for *girl*, work out the components of the meanings of the following lexemes:

[1] boy   woman   child   adult   human

For *boy* the components will be [+HUMAN], [−ADULT], [+MALE]; for *woman* [+HUMAN], [+ADULT], [−MALE]. For *child* there will be only two components: [+HUMAN], [−ADULT]; similarly for *adult*: [+HUMAN], [+ADULT]; these two lexemes are not distinguished for sex. In the case of *human* there will be only one component, [+HUMAN], since this is not distinguished either for sex or for maturity.

What has emerged from our discussion so far is that components have a distinguishing function. They serve to distinguish the meaning of a lexeme from that of semantically related lexemes, or more accurately they serve to distinguish among the meanings of lexemes in the same semantic domain. We can display the semantic distinctions by means of a matrix:

[2]	[MALE]	[ADULT]
man	+	+
woman	−	+
boy	+	−
girl	−	−

This shows that the semantic components [MALE] and [ADULT] serve to distinguish the meanings of these four lexemes.

We could perhaps identify components for distinguishing the meanings of lexemes that are semantically unrelated, say *girl* and *cup*, e.g. [+ANIMATE] vs. [−ANIMATE]; but it is questionable whether identifying components at this level of generality tells us much of interest about the meanings of lexemes. We shall therefore concentrate on semantic components as distinguishing features of lexemes within a specified semantic domain (compare Chapter 14).

The semantic domain where componential analysis was first used with some success was kinship terminology (e.g. Lounsbury, 1964). How many semantic components do you think are needed to distinguish the following kinship terms of English? *Note*: kinship terms are conventionally described in relation to a given person, technically termed by the Latin equivalent of the pronoun *I*: *ego*.

[3] mother   father   uncle   aunt   brother   sister
    daughter   son   nephew   niece   cousin

We shall clearly need a semantic component to distinguish the gender of the lexemes, i.e. [MALE]; though the lexeme *cousin* is not distinguished for gender and would therefore be marked [+/− MALE]. Secondly, we need a semantic component to make distinctions of generation (in respect of *ego*); e.g. *brother* and *sister* are the same generation as *ego*, while *father* and *mother* are one generation above (ascending generation) and *son* and *daughter* are one generation below (descending generation). We therefore need two semantic components to distinguish the generations: [ASCENDING] and [DESCENDING]. Let us now construct a matrix to see how far we have come in our analysis of these kinship terms.

[4]	[MALE]	[ASCEND]	[DESCEND]
father	+	+	−
mother	−	+	−
uncle	+	+	−
aunt	−	+	−
brother	+	−	−
sister	−	−	−
son	+	−	+
daughter	−	−	+
nephew	+	−	+
niece	−	−	+
cousin	+/−	−	−

We have not yet fully distinguished the meanings of these kinship terms in English, since *father* and *uncle* have the same analysis, as do *mother* and *aunt*, *son* and *nephew*, and *daughter* and *niece*. We therefore need a component that is common to *father*, *mother*, *son* and *daughter* as against *uncle*, *aunt*, *nephew* and *niece*. Here then we are talking about 'direct' or 'lineal' descent as against 'collateral' descent, and we might propose a semantic component of [LINEAL]. Our matrix in [4] will now be modified to that in [5] on page 83.

We now have a unique analysis for each term in the kinship system, and we have distinguished the eleven terms with binary choices from four semantic components.

[5]

	[MALE]	[ASCEND]	[DESCEND]	[LINEAL]
father	+	+	−	+
mother	−	+	−	+
uncle	+	+	−	−
aunt	−	+	−	−
brother	+	−	−	+
sister	−	−	−	+
son	+	−	+	+
daughter	−	−	+	+
nephew	+	−	+	−
niece	−	−	+	−
cousin	+/−	−	−	−

Types of component

We have stated that the componential analysis of meanings is particularly applicable to distinguishing the meanings of lexemes that are semantically related (in the same semantic domain). This suggests that there are two broad types of semantic component: those that serve to identify a semantic domain and that are shared by all the lexemes in the domain; and those that serve to distinguish lexemes from each other within a semantic domain. The first type of component is sometimes called a **common** component, and the second type is termed **diagnostic** or, as in phonology, a **distinctive feature**. For example, the meanings of the lexemes in the limited domain of *man, woman, boy* and *girl* have the common semantic component [HUMAN] and the diagnostic components [ADULT] and [MALE]. If we were to expand the domain to include all lexemes referring to mammals, then [HUMAN] would become a diagnostic component. We might note at this point that some semantic components can be said to presuppose other, more general, semantic components: [+HUMAN], for example, presupposes [+MAMMAL], which in turn presupposes [+ANIMATE]; but we cite·such components only if they have a function in defining a semantic domain (as a common component) or in distinguishing the meanings of lexemes (as a diagnostic component). The lexemes in [3] represent a semantic domain defined by the common components [HUMAN], [KINSHIP] (note that *man, woman*, etc. are [−KINSHIP]) and distinguished by the

diagnostic components [MALE], [ASCENDING GENER-
ATION], [DESCENDING GENERATION] and [LINEAL
DESCENT].

Consider now the lexemes in [6], which all refer to
kinds of fastener. What diagnostic components would serve
to distinguish the meanings of these items?

[6] screw   nail   rivet   tack

A *screw* is distinguished from the others by possessing a
threaded shank, requiring it to be turned (with a screwdriver) rather than driven (with a hammer): we might
propose a component [THREADED] or a component
[DRIVEN] to represent this distinction. A *rivet* is
distinguished from the others by possessing a flat end rather
than a pointed end: a component [POINTED] would
represent this distinction. We now need a component to
distinguish *nail* and *tack*. We might consider length to be
a distinguishing feature: tacks are generally relatively short
by comparison with nails, though nails (and screws) come
in a variety of lengths, and rivets are generally short,
though length is hardly a distinctive feature for them.
Alternatively, we might consider more significant the fact
that tacks are usually used to fix fabrics (including carpets),
while nails are more often used for fixing wood-type
materials, and so propose a component [FABRIC]. Let us
now construct a matrix to test the diagnostic components
we have proposed:

[7]	[THREADED]	[POINTED]	[FABRIC]
screw	+	+	−
nail	−	+	−
rivet	−	−	−
tack	−	+	+

None of these items has the same analysis; the components
serve to distinguish the meanings of these lexemes. Our
analysis may be challenged however if we begin to expand
the domain to include further kinds of fastener, e.g. *bolt*,
pin. The analysis is also suspect because of the *ad hoc* nature
of, for example, the component [THREADED]: it is a
component proposed merely for this limited analysis and is
unlikely to be generalisable to other analyses. We begin to

see the limitations of componential analysis.

For the moment let us note that the diagnostic components used in [7] are of two types. The first two, [THREADED] and [POINTED] are **formal** components: they relate to features of the form of the objects denoted by the lexemes in [6], i.e. whether they are provided with a thread or whether they have a point. The third component, [FABRIC], on the other hand, is **functional**: it relates to what the object is used for, what its function is. In general, formal features seem to be preferred to functional ones in making distinctions of meaning among lexemes referring to objects. If you recall, we could have used the formal component of length to distinguish *tack* from the other fasteners, though it seemed doubtful whether it would make the distinction between *tack* and *nail* positively enough.

Consider now the lexemes in [8], which refer to different kinds of seats. Is it possible to distinguish their meanings by just formal diagnostic components?

[8] chair   stool   sofa   bench

Stools and benches are distinguished from chairs and sofas in that the former do not usually have backs: we can propose a formal component [+/− BACK]. Sofas are always upholstered, chairs may be, stools and benches are usually not: the formal component [+/− UPHOLSTERED] will distinguish *sofa* and *chair*. How then do we distinguish *stool* and *bench?* In terms of form we might invoke a component of length, i.e. [+/− LONG], which we might also use to distinguish *sofa* and *chair* (and so ignore the upholstery). Length correlates with the functional component 'for two or more persons', which we would accept intuitively more readily than 'long' vs. 'short'. However, we have to say that it would be possible to use just formal components in distinguishing the meanings of the lexemes in [8], as follows:

[9]	[BACK]	[LONG]
chair	+	−
stool	−	−
sofa	+	+
bench	−	+

If we were now to add the item *high chair* to the lexemes in [8] we would find it difficult to avoid invoking a functional component, since the criterial feature of a high chair is that it is 'for young children'. Again, we may note that the components proposed do not have a very wide range of applicability, by contrast with components like [CONCRETE] or [HUMAN].

All the semantic components that we have proposed so far have represented binary choices: either the component is present in the meaning of a lexeme or it is not. For some proponents of componential analysis the binary nature of semantic components is axiomatic. Before we question this point, let us note that a binary component in fact allows three possibilities: either it is present [+], or it is absent [−], or it may be present or absent [+/−]. We used this third possibility in [4] in relation to the component [MALE] and the lexeme *cousin*. Alternatively, we might say that the component [MALE] is irrelevant to the meaning of *cousin*, and we would mark it as [0].

Look back now to [4]. In dealing with the aspect of the meaning of kinship terms relating to generation we needed to make a three-way distinction: the same generation as *ego*, the generation above *ego* (ascending), and the generation below *ego* (descending). In order to make that distinction we invoked two binary components, [+/− ASCENDING] and [+/− DESCENDING]. There is a certain redundancy here, since two binary components are capable of making a four-way distinction, as in [2] and [9]. However, these generation components are different from the [ADULT] and [MALE] ones: the minus value in each case implies the other one and an intermediate value ('the same'), e.g. [−DESCENDING] implies both [+ASCENDING] and 'same'. Consequently, 'the same generation' is marked as [−DESCENDING], [−ASCENDING]. The difficulties are compounded if we introduce terms into the kinship system referring to two generations above or below *ego*, e.g. *grandfather, grandmother, grandson, granddaughter*. With binary components we would need an [ASCENDING2] and a [DESCENDING2] component, and again the minus value would imply everything else.

These difficulties could be overcome by allowing nonbinary components, i.e. components with multiple values.

For example in [4] we could replace the components [ASCENDING] and [DESCENDING] by a single component [GENERATION], which would have the values '0' for same generation, '+1' for one generation above *ego*, '−1' for one generation below *ego*: and then we could add '+2' and '−2' for grandparents and grandchildren respectively. Some components would continue to be binary, e.g. [MALE]; a person is normally either male or female. Similarly, descent is either lineal or collateral. However, we might wish to be uniform and propose components with two values; e.g. [GENDER] would have the values 'male' and 'female', [DESCENT] the values 'lineal' and 'collateral'. In the case of the [GENDER] component, we may consider this to be an improvement over [+/− MALE], in order to avoid the sexism of [− MALE]. We could rewrite [5] as [10]:

[10]

	[GENDER]	[GEN'TION]	[DESCENT]
father	m	+1	l
mother	f	+1	l
uncle	m	+1	c
aunt	f	+1	c
brother	m	0	l
sister	f	0	l
son	m	−1	l
daughter	f	−1	l
nephew	m	−1	c
niece	f	−1	c
cousin	m/f	0	c

Now propose diagnostic semantic components for the lexemes in [11]. Take the lexemes to be referring only to movement by human beings. Attempt binary components first and then non-binary components.

[11] walk   run   crawl   jog   sprint

The meanings of these words have the common components of movement across a solid surface (contrast *swim*, *fly*). The crucial differences are the number of limbs in contact with the surface (four in *crawl*, two in the others), whether at least one limb remains in contact with the surface at all times (no in the case of *run, jog, sprint*, and yes for the other two), and the speed of the movement. For the first of these differences we need two binary components [+/− FEET], [+/− HANDS]. For the second we need a component [+/− CONTACT]. And for the third, we need to distinguish three speeds (for *jog, run and sprint*), for which we will need two binary components, [+/− FAST] and [+/− SLOW]. Our matrix will be constructed as follows:

[12]   [FEET] [HANDS] [CONTACT] [SLOW][FAST]

	[FEET]	[HANDS]	[CONTACT]	[SLOW]	[FAST]
walk	+	−	+	+	−
run	+	−	−	−	−
crawl	+	+	+	+	−
jog	+	−	−	+	−
sprint	+	−	−	−	+

Arguably with *crawl* it is not so much the feet that are in contact with the surface as well as the hands, but rather the knees or indeed the whole lower leg. This may be expressed more easily with a non-binary component of [LIMBS] to replace [+/− FEET] and [+/− HANDS]. Similarly the two binary components of speed may be replaced by a single non-binary component of [SPEED], thus removing the anomaly where *run* is marked as [−SLOW] and [−FAST]. Our matrix may now be constructed as follows; though this is not the only possible solution:

[13]   [LIMBS]   [CONTACT]   [SPEED]

	[LIMBS]	[CONTACT]	[SPEED]
walk	feet	+	slow
run	feet	−	fast
crawl	hands+legs	+	slow
jog	feet	−	slow
sprint	feet	−	very fast

Let us now add to the lexemes in [11] the items *stroll* and *saunter*. Can you suggest how we might distinguish the meanings of these from that of *walk*?

This is rather difficult. We might think that *stroll* and *saunter* involve a slower movement than *walk*, but this is hardly a crucial distinction. The *Longman Concise English Dictionary* defines *saunter* as 'to walk about in a casual manner' and *stroll* as 'to walk in a leisurely or idle manner'. It is the element of casualness that characterises *saunter*, and leisure-liness or idleness *stroll*. How do we capture these meanings in semantic components, either binary or non-binary? Perhaps we can do it with a component of [MANNER]. This question, to which there is, I believe, no entirely satisfactory answer, leads us on to ask about the extent to which componential analysis is applicable to the description of meaning.

Applicability

The difficulty with *saunter* and *stroll* is a problem of **meta-language**. Metalanguage is language used to talk about language. Linguistics is unique among the sciences, natural and social, in having as its object of study the same phenomenon that provides the means of description: language. As in so many other sciences, linguists have coined their own technical terms – phoneme, morpheme, lexeme – but when it comes to the description of the meanings of words it would be impossible to propose a set of technical terms to comprise the metalanguage of semantics. And yet, in a way, that is what components are intended to be, though they are, as are many terms of description, borrowed from everyday language. We have illustrated the point with *saunter* and *stroll*. Consider now the items in [14]. Can you suggest diagnostic components to distinguish the meanings of these lexemes?

[14] acquaintance   friend   colleague   associate

All these items have a common semantic component of 'relationship between persons, usually positive'. *Colleague* and *associate* refer to work or business relationships, whereas *acquaintance* and *friend* are not marked in this way. On the other hand, they could not be said to have a semantic component [−WORK]: we can talk of 'a friend from work' or 'a business acquaintance'. *Acquaintance* and *friend* differ perhaps in the closeness of the relationship, and we might propose a component [+/− CLOSE] to distinguish them. But this component does not seem to be of relevance to *colleague* and *associate*, where the distinction seems to revolve around the contexts in which we might use these items.

Associate is generally more applicable to a business context, while *colleague* is associated with a professional context; but there is a certain amount of interchange possible. Again, a semantic component of 'context' is not relevant to *acquaintance* and *friend*. When we begin to analyse the meanings of lexemes referring to relationships it is not nearly so easy to propose semantic components as it is when analysing the meanings of lexemes which refer to objects, where a metalanguage of formal properties provides a suitable means of description. Neither would we necessarily expect to find agreement among analysts or native speakers about what is criterial to meaning differences between more abstract lexemes.

Is there then a set of semantic components which is universal and from which the meanings of lexemes in all languages are composed? If there is, we do not yet have the knowledge or the metalanguage to specify what such a set might be. While there is much that is common to human experience the world over and which human beings want to talk about, making translation between languages a real possibility, different cultures nevertheless often have vastly different organisations, sets of artefacts, institutions and norms of behaviour. Consequently, the meaning distinctions that are relevant to one culture (and the semantic components necessary for describing them) may not fit another culture at all. For example, all cultures have kinship systems, but they are often organised in a quite different way, e.g. in the cultures of American Indians or Australian Aboriginals. While components like [+/− ADULT] and [+/− MALE] may feature in the analysis of all of them, there will be many distinctions that are culturally specific, e.g. where a distinction is made lexically between brothers and sisters older and younger than *ego*.

We may conclude therefore that there is no universal set of semantic components from which the meanings of lexemes are composed. Indeed we may question whether the meaning of a lexeme can be said to be the sum of its semantic components. Is the meaning of *girl* adequately described as [+HUMAN], [−ADULT], [−MALE]? Or can we say that the meaning of *screw* is [+FASTENER], [+THREADED], [+POINTED], [−FABRIC]? We need to remind ourselves of the reasons for our selection of semantic components: we selected them in order to discriminate between the meanings of lexemes in the same semantic domain. There was no suggestion that we would thereby fully characterise the meanings of the lexemes concerned. Indeed it has been our position in this book so far that the description of the meaning of a lexeme must involve a number of perspectives, e.g. denotation, sense relations and, as we shall see in the next chapter, collocation. At the same time, though, componential analysis is limited in its range of applicability: it does not apply easily to all areas of the vocabulary. Semantic components, when they can be identified, have a discriminatory function; they add to our understanding of the meaning of a lexeme by providing points of contrast with semantically related lexemes.

By the same token, componential analysis can add to our understanding of synonymy (see Chapter 5). A pair of true synonyms will share the same set of semantic components. For example, *drawing pin* and *thumb tack* share the components say [+FASTENER], [+POINTED], [+BROAD HEAD], [+FOR PAPER]; so that the difference between these lexemes is clearly located elsewhere, in this case in dialect, with the former belonging to British English and the latter to North American English. A componential analysis may also help us to establish degrees of synonymy. We may talk of a looser synonymy where a pair of lexemes has some but not all semantic components in common. For example, *barn* and *shed* would be looser synonyms. They share the components [BUILDING], [STORAGE], but *barn* has the additional component of [FARM] and perhaps that of [FOR CEREALS], while *shed* has perhaps the additional component [HOUSE]. In the case of antonymy, a pair of antonyms usually share all their components except one; e.g. *man* and *woman* share the components [+CONCRETE], [+ANIMATE], [+HUMAN], but they are contrasted by the component [MALE], *man* having the [+MALE] component, *woman* having the component [−MALE]. Clearly the same limitations apply to this application of componential analysis as generally: it works best with sets of lexemes referring to objects.

Likewise componential analysis may help us in understanding the sense relation of hyponymy, which we described briefly in Chapter 5 (p. 64) and which we shall mention again in Chapter 14 (p. 213). Hyponymy refers to the relation of inclusion of meaning, e.g. the fact that the meaning of *rat* is included in the meaning of *rodent*. Included meanings will share the semantic components of more general meanings, while the latter will not have the specific components of the included items. For example, the meanings of *man, woman, boy* and *girl* are included in the meaning of *human being*: they share with it the semantic component [+HUMAN], but it does not share the components [+/− ADULT] and [+/− MALE], needed to discriminate them. Similarly, the meanings of *man* and *woman* are included in the meaning of *adult* (the shared component is [+ADULT]), and the meanings of *boy* and *girl* are included in the meaning of *child* (with the shared component [−ADULT]). *Boy* and *girl* are then hyponyms of *child; man* and *woman* are hyponyms of *adult*; and *child* and *adult* in turn are hyponyms of *human being*.

We have been emphasising in this section that componential analysis has a useful part to play in contributing to the description of the meanings of lexemes, but that there are limitations on its applicability. It works best with taxonomies (systems of classification, e.g. kinship) or sets of concrete objects. It is of more doubtful value in describing the meanings of more abstract lexemes, not least because we lack an adequate metalanguage. Consider the set of lexemes: *annoy, irritate, vex, displease, provoke*. They all refer to ways of causing someone to be angry or to feel angry; any member of the set is frequently defined in terms of one or more of the other members. In attempting a componential analysis, once we have got beyond the common component [CAUSE ANGER], which is itself rather *ad hoc*, it is difficult to imagine what terms we might propose for distinguishing the meanings of these lexemes. Think, for example, how you might explain the differences in meaning to a child or a foreign learner. You would most likely go straight to examples of typical collocations (see Chapter 7, p. 96). There is no terminology available for describing the meanings of such sets of abstract lexemes. Let us now see whether lexicographers use componential analysis in writing dictionary definitions of lexemes.

Components and definitions

In a way the alphabetical organisation of dictionaries mili-
tates against a systematic use of componential analysis in the
composition of definitions. Semantically related lexemes are
not found grouped together in a dictionary as it is traditionally conceived (though see Chapter 14, p. 216); each
lexeme is treated individually; there appears to be no
compelling reason to concentrate on the discriminatory
components of the meanings of lexemes. Nevertheless,
meaning implies contrast, and the description of meaning
implies drawing contrasts with lexemes that are semantically contiguous. The problem is rather with the alphabetical organisation, as we shall discuss in Chapter 14. Look
up the definitions of the lexemes in [15] in your dictionary.
Is there any sense in which the definitions use semantic
components?

> [15] crepe   seersucker   tweed   velvet

In the *Longman Concise English Dictionary* definitions of these
lexemes, a common semantic component is identified by the
use of the word *fabric*. Then, if we compare the definitions
further, reproduced at [16], we can note a number of
repeated types of feature:

[16]	*crepe*	'a light crinkled fabric woven from any of various fibres'
	seersucker	'a light slightly puckered fabric of linen, cotton or rayon'
	tweed	'a rough woollen fabric made usu in twill weaves and used esp for suits and coats'
	velvet	'a fabric (e.g. of silk, rayon or cotton) characterised by a short soft dense pile'

In each of these definitions there is a value for a non-binary
component of [FIBRE]: for *crepe* it is 'various', for *seersucker*
it is 'linen/cotton/rayon', for *tweed* it is 'wool', for *velvet* it
is 'silk/rayon/cotton'. Additionally there is a value for a non-
binary component [TEXTURE]: for *crepe* 'crinkled', for

seersucker 'slightly puckered', for *tweed* 'rough', and for *velvet* 'with short soft dense pile'. It is the texture that is especially characteristic of *velvet*. The definitions of the first two lexemes of [16] have a component that relates to their weight: this is perhaps a binary component, with a minus value in these two cases, i.e. [−HEAVY]. The definition of *tweed* contains additionally information about its method of production ('made in twill weaves') and its function ('used esp. for suits and coats'): these, too, might be regarded as semantic components that may be invoked in discriminating the meanings of lexemes in the semantic domain of 'fabric'.

Now look up the lexemes in [17] in your dictionary. Is it possible to identify the use of semantic components in their definitions?

    [17] approve   confirm   ratify   sanction (verb)

The relevant definitions from the *Longman Concise English Dictionary* are given in [18]:

    [18] *approve*   'to give formal or official sanction to; ratify'

         confirm   'to give approval to; ratify'

         ratify   'to approve or confirm formally'

         sanction 'to make valid; ratify'

You will notice that these definitions depend largely on the use of synonymy, together with a measure of paraphrase (see Chapter 9, p. 135). A componential analysis appears not to have informed the definitions in these cases. But this is not surprising. We are here concerned with a set of verbs with a fairly abstract meaning, and we have noted that componential analysis does not apply easily to abstract lexemes. The dictionary definitions only reflect our conclusions about componential analysis.

Exercises

1. Make a componential analysis, presented in the form of a matrix, of the following lexemes, using only binary components:

    bitch   *cat*   *dog*   *kitten*   *puppy*   *tomcat*

When you have done that, try extending your analysis by adding further lexemes of a similar kind, e.g. *horse, mare, foal.*

2. Make a componential analysis as in 1. of the following lexemes, but use non-binary components as well if necessary:

      cup goblet mug tankard tumbler

   You can try extending your analysis by adding further semantically related lexemes, e.g. *glass.*

3. Make a componential analysis as in 2. of the following lexemes. What limitations does componential analysis have as a method for describing the meanings of words?

      clarinet cymbal harp trumpet violin

4. Look up the following lexemes in your dictionary. For which of them do you think the definitions give evidence of componential analysis?

      alligator deprive finale herald jerkin litotes nonchalant pewter rondeau tortuous

CHAPTER 7
Meaning from Combinations

The perspectives on meaning that we have explored so far include the referential relation between a word and an entity in the world, and sense relations between words within the structure of vocabulary. Componential analysis applied essentially to referential meaning, though with some relevance to sense relations, e.g. in the explanation of synonymy, antonymy and hyponymy. We have been viewing words by and large either as individual items or as substitutes for one another. In this chapter we turn our attention to the lexical and semantic relations that a word has with other words that accompany it in the stream of speech or writing: its **syntagmatic** lexical relations. This is a different study from that of syntax in grammar: grammatical syntax is about the structure of sentences in terms of classes of words (e.g. nouns as a class, verbs as a class) and their combinations. Syntagmatic lexical relations (i.e. arising from combination in structures like phrases and sentences) are concerned with individual lexemes and the meaning relations they enter into with other accompanying lexemes. We begin by looking at the meanings that arise from what is called **collocation**.

Collocational meaning

Collocation refers to the combination of words that have a certain mutual expectancy. The combination is not a fixed expression (see below, p. 103), but there is a greater than chance likelihood that the words will cooccur. For example, let me ask you to supply a list of nouns that might follow *false* in the following sentence:

[1] He had a false _____.

The nouns that would spring readily to mind to fill the slot in this structure might include: *eye, nose, beard, expectation, passport/identity paper.* If the indefinite article (*a*) were omitted, we might add *teeth, eyebrows.* If the subject of the sentence were *the car* rather than *he*, then we might expect *numberplate.* We can note that while we talk of *false* eyes, noses and teeth, we tend to use *artificial* with arms and legs. What these examples illustrate is that words regularly keep company with certain other words, and it is such combinations that we refer to as 'collocations'.

We would go further than this and say that the collocations that a lexeme regularly enters is a factor that needs to be taken account of in the description of its meaning. Part of the meaning of *false* therefore is the fact that it is regularly found in combination with *teeth, eye, expectation, passport,* etc. Since collocation involves mutual expectancy, then part of the meaning of *tooth* is that it combines regularly with *false*, among other lexemes. However, the strength of the expectancy may not be equal in both directions: *tooth* is probably more likely to occur in combination with *false* than *false* is to occur in combination with *tooth.* Similarly, with the collocation *good read*, as in 'Dickens' latest novel is a good read', it is more likely that, given the noun *read*, the adjective will occur, than, given the adjective *good*, the noun *read* will occur. That is to say, the number of alternative adjectives to *good* in the phrase *good read* is relatively limited by comparison with the number of alternative nouns to *read.* Indeed in this case we might say that its likely combination with *good* is part of the meaning of *read* (noun), but we would be far less sure about saying that its combination with *read* (noun) is part of the meaning of *good.*

There are a number of ways in which collocation is relevant to the description of the meaning of a lexeme. Consider the collocations of the lexeme *strong* given in [2]. What would you say the meaning of strong is in each case?

[2] a strong woman   a strong door   strong tea   a
    strong personality

In *a strong woman* the word *strong* is referring to physical strength and the ability to perform actions requiring physical strength. In *a strong door* the reference is also to physical strength, but in a passive sense: a strong door is one that is not easily broken down. In *strong tea* a quite different meaning of *strong* is present: here the reference is to the intensity of the flavour and perhaps also to the darker colour of the tea. Finally, in *a strong personality* the strength referred to is not physical but rather moral and implies that a person is influential or persuasive. In each of these combinations, then, *strong* has a different meaning (or sense), and if you look up *strong* in a desk-size dictionary you will find these meanings listed, among others, as different senses of the word. These different senses of *strong* arise in large part from the specific collocations of *strong*. It is the collocation that determines which particular sense of *strong* is meant. These collocations are so regular that they are recognised by lexicographers as providing different senses of the lexemes. Collocations, then, may lend specific meanings to a lexeme.

Consider now the verb *hiss*. Which nouns as subject would you readily associate with *hiss*, e.g. in the sentence 'The _____ is hissing'?

There are probably three, fairly limited, groups of nouns that regularly combine as subject with *hiss*. Firstly, there are living creatures, especially *snake* and *cat*. Secondly, there are collections of people, e.g. *audience, crowd*, which hiss to show disapproval. And thirdly, there are cooking utensils that emit steam with a hissing sound, perhaps *kettle* and *pressure cooker*. The meaning of *hiss* does not differ much in collocation with any of these groups of nouns: the same kind of sound is being referred to in all cases. What we have identified are the nouns that regularly cooccur as subject with *hiss*. This kind of meaning relation is often referred to as the collocational **range** of a lexeme. The range of *hiss* then includes certain creatures, collections of (disapproving) people, and certain cooking utensils. Arguably, this kind of information contributes to a description of the meaning of *hiss*. It certainly helps us to account for an unusual collocation of *hiss*, when the range of *hiss* is extended, e.g. when a novelist writes

[3] "I'll get you back for this," she hissed.

Collocational range then contributes to the meaning of a lexeme and helps to explain range extending tendencies of lexemes. It also helps us to interpret metaphor in poetry. For example, when Ted Hughes in 'October Dawn' writes 'premonition of ice', we interpret it on the basis of our knowledge of the typical collocation of *premonition* with *death* or *disaster*.

Now consider the adjective *rancid*. Which nouns would you use in combination with *rancid*, e.g. in the sentence 'The _____ is rancid'?

The noun that immediately came to your mind was no doubt *butter*. You may have also associated *bacon* with *rancid*, but it is doubtful whether you have been able to propose any more nouns beyond these. Even though whenever *rancid* occurs it is almost always in the company of *butter*, the combination *rancid butter* or *butter (be) rancid* is not a fixed expression, since *butter* may occur with several other adjectives. What we have is an item (*rancid*) with an extremely restricted range. We might speak more accurately of a **collocational restriction**. Clearly, the meaning of *rancid* is heavily dependent on the fact that it is collocationally restricted to *butter*, and perhaps additionally to *bacon*.

Discovering collocational patterns

All the examples of collocations that we have discussed so far we have so to speak thought up out of our heads, drawing on our knowledge and experience of using the language. This is really a rather unsatisfactory way of obtaining data on collocational patterning. Indeed we may argue that, in spite of its widespread practice by linguists, it is an unsatisfactory way of arriving at any linguistic data. It is preferable to observe language in use, in the everyday spoken and written communication of language users, as the basis of the study of linguistic phenomena. In order to proceed in this manner, linguists usually collect a corpus of spoken and written texts which will be as representative of the range of language varieties as is appropriate to the linguistic investigation being undertaken.

The first step in discovering the collocational patterns of English is to provide ourselves with a corpus consisting of a representative sample of texts. We then need to decide which lexemes we are intending to investigate in order to discover the other lexemes that regularly cooccur with it. The lexeme under investigation is called the **node**. We have to decide how many words either side of the node we are going to look at in order to find regularly cooccurring lexemes. This is called the **span**. A span of five words either side of the node is usually sufficient to throw up all the significant collocations of that node. The next part of the task is basically a counting one. We need to go through our corpus and find all the examples of the node (lexeme) that we are interested in. Then we count how many times each of the five words to the left and each of the five words to the right of each occurrence of the node occurs in the corpus as a whole. We can then construct a table of lexemes that occur in the company of the lexemes that we are interested in (its **collocates**) and how often they occur. Clearly, the more frequently they occur, and if they occur with a frequency greater than chance, then we will have identified a significant collocational pattern. Needless to say there will be many collocates that occur only once in the span of the node and will thus not be of collocational significance.

Let us attempt a small-scale investigation on the text in [4]. We will choose the lexeme *princess* as our node. Count the collocates within a span of five and list them in order of frequency of occurrence.

[4] Once upon a time there was a prince who wanted to marry a princess, but she would have to be a real princess. He travelled around the whole world looking for her; but every time he met a princess there was always something amiss. There were plenty of princesses but not one of them was quite to his taste. Something was always the matter: they just weren't real princesses. So he returned home very sad and sorry, for he had set his heart on marrying a real princess.

One evening a storm broke over the kingdom. The lightning flashed, the thunder roared, and the rain came down in bucketfuls. In the midst of this horrible storm, someone knocked on the city gate; and the king himself went down to open it.

On the other side of the gate stood a princess. But goodness, how wet she was! Water ran down her hair and her clothes in streams. It flowed in through the heels of her shoes and out through the toes. But she said that she was a real princess.

"We'll find that out quickly enough," thought the old queen, but she didn't say a word out loud. She hurried to the guest room and took all the bedclothes off the bed; then on the bare bedstead she put a pea. On top of the pea she put twenty mattresses; and on top of the mattresses, twenty eiderdown quilts. That was the bed on which the princess had to sleep.

In the morning, when someone asked her how she had slept, she replied, "Oh, just wretchedly! I didn't close my eyes once, the whole night through. God knows what was in that bed, but it was something hard, and I am black and blue all over."

Now they knew that she was a real princess, since she had felt the pea that was lying on the bedstead through twenty mattresses and twenty eiderdown quilts. Only a real princess could be so sensitive!

The prince married the princess. The pea was exhibited in the royal museum, and you can go there and see it, if it hasn't been stolen. Now that was a real story.

(Hans Andersen: 'The Princess and the Pea')

When you undertake an exercise like this you realise how important it is to bear in mind the discussion that we had in Chapter 1 about what we mean by 'word'. On the assumption that you have done the same, I have counted orthographic words. It will serve for purposes of illustration, though it is not really adequate for a collocational analysis. The results of the count are as follows:

NO. OF OCCURRENCES	WORDS
10	the
9	a
8	be (inc. was, were)
6	real
5	she
4	have (inc. had), to
3	but, he, marry (inc. marrying, married), of, that
2	amiss, in, not, on, one, sensitive, so, there, will/would
1	always, around, bed, broke, could, eiderdown, evening, every, exhibited, felt, find, gate, goodness, heart, home, how, just, met, only, out, pea, plenty, prince, quilts, returned, since, sleep, something, stood, storm, then, they, time, travelled, very, wanted, we, wet, which, who, whole

There are twelve occurrences of *princess* in [4], giving us a total of 120 collocates in a span of five. The definite and indefinite articles are the most frequently occurring items in the company of *princesss*, but they hardly constitute a significant collocation. They are likely to appear so frequently that their occurrence is of no significance. This is true of all the grammatical words (see Chapter 1, p. 15), which we should therefore exclude from our collocational study. We also exclude words which occur only once, on the basis that a single occurrence is not sufficient to constitute a significant collocation. If we do that we identify the following lexical words with a frequency of occurrence greater than one:

[5] real (6) marry (3) amiss (2) sensitive (2)

The item *amiss* is rather odd. It occurs only once in the text, but it falls in the span of two occurrences of *princess*. So we should probably discount that item also, giving us a list from this brief sample text of three items having a significant collocation with *princess*:

[6] real   marry   sensitive

This is only a short text (less than four hundred words) and hardly a representative sample of English. We can hardly generalise our results to the language as a whole or make any reliable statement about the collocation of *princess* on the basis of it (it is surprising, for example, that *beautiful* did not occur), but it has served to illustrate what is involved in collocational analysis. You no doubt found it a fairly tedious exercise looking for the collocations of one item in a short text. Think of what would be involved in discovering the collocations of a large number of items in a representative sample of several million words – which is what would be needed to gain significant results. Perhaps this is why so few collocational studies have been undertaken. These days, however, much of the drudgery can be taken out of the task by using computers, though it is still a considerable undertaking, and until very recently a large enough corpus of texts has not been available on computer.

Lexicographers, therefore, and others who wish to take account of collocation in the description of meaning, have largely been left with their intuitions and insights into their own and fellow-speakers' knowledge of language use. Far more easy to identify are combinations of words that always occur in a more or less fixed form, and to these we now turn our attention.

Fixed expressions

Collocations vary in the degree to which one lexeme expects another to occur with it. We have seen that *rancid* has a very strong expectation for one or two cooccurring lexemes (*butter, bacon*). In the case of others, where the choice is greater (e.g. *hiss*), the expectation is less strong; and in some cases (e.g. general evaluative adjectives like *good*) expectation of cooccurrence is quite weak. 'Rancid butter' almost constitutes a fixed expression, but collocation is by definition not fixed, since there is always some degree of choice.

The fixed expressions that we are going to discuss in the rest of this chapter are: cliches, proverbs and idioms. But we need first of all to distinguish these from some other fixed-word combinations that are not fixed expressions as such. The combinations concerned are some of the multi-word lexemes that we discussed in Chapter 1, viz. phrasal and prepositional verbs, and compound lexemes that happen not to be written together. So, items like *give up* (phrasal verb), *look after* (prepositional verb), *put up with* (phrasal-prepositional verb), and *child minder* (compound) we are excluding from the category of fixed expression.

Cliches might be regarded as kinds of ossified collocations. In certain contexts the mutual expectancy of lexemes has become fixed. The result is a loss of meaning, because there is no longer an element of choice or contrast. For example, we ridicule estate agents' advertisements for describing houses as 'desirable residences'. The lexeme *residence*, itself a high style synonym for *house*, seems to be always accompanied by *desirable* in a fixed expression, and we read it no longer as having the meaning 'desirable' + 'residence', but merely as a paraphrase for *house*: it has become a cliche of estate agents' jargon. Similarly in postal advertising in particular the noun *owner* is invariably accompanied by the adjective *proud*: 'You could be the proud owner of a . . .'.

Consider now the nouns in [7]. The first three come from the context of advertising and the remainder from the context of negotiation. What adjective do you most readily associate with each of them?

[7] bargain offer (price) – reduction decisions discussions precedent progress refusal

---

A number of adjectives may suggest themselves for *bargain*, all of them more or less cliched: *real, genuine, fantastic*. Apart from a kind of general intensification they add little to the meaning of *bargain*. *Real* is used with a number of nouns in a rather cliched way, e.g. *meaning, possibility*. An *offer* is typically *unbeatable*, and a *price reduction* usually *genuine*. In negotiating jargon we usually hear that *decisions* are *difficult*, *discussions* are *useful*, *precedents* are *dangerous*, *progress* is *real* (again), and *refusals* are *flat*.

We do not usually find cliches obtrusive. We seem to be able to filter out semantically empty expressions in our reading and listening. If we do become aware of cliches, though, we may find them rather annoying, because we may regard them as a waste of words and a semantic devaluation. More obvious in speech and writing are **proverbs** because there is an incongruity between the literal meaning of a proverb and the context to which it refers. When we say 'You can't have your cake and eat it' we are not usually referring to literal cake. We are using the proverb as a graphic, though perhaps less direct, way of saying to someone that they have to choose one of two options, they cannot have the advantages of choosing both.

Proverbs represent a common cultural fund of folk knowledge and wisdom that we can call on to warn or reprimand someone in the assurance that they will accept the basis of this common wisdom where a more direct personal approach would fail. Or we use proverbs to comment on or come to terms with life's experiences; e.g. 'You can't win them all', said as a consolation following a disappointment or failure. Consider the proverbs in [8]. On what kinds of occasion would you expect to hear them used?

[8a] A bird in the hand is worth two in the bush.
[8b] People who live in glass houses shouldn't throw stones.
[8c] A stitch in time saves nine.
[8d] A new broom sweeps clean.
[8e] Too many cooks spoil the broth.

The proverb in [8a] is used when we may have the opportunity of some advantage immediately but might prefer to wait for a supposed greater though by no means certain advantage in the future. We use the proverb in [8b] of someone who supposedly suffers from the same fault that they are criticising in another. We use [8c] to persuade someone to take immediate action before procrastination may mean that a task will be more difficult and complex than it is now. When a new person is appointed to a senior job in an institution and begins to make major changes we may ruefully utter [8d]. And [8e] is used when too many people try to do a particular job or try to solve a particular problem. Proverbs are sometimes alluded to by quoting a part of them, e.g. 'a case of too many cooks' or 'you're counting your chickens again'; and these abbreviated forms then themselves become idioms. Proverbs are probably less current in everyday conversation than they were a generation or so ago; perhaps a sceptical generation no longer plugs in to the accumulated folk wisdom of a culture.

Some lexicologists would regard proverbs as a kind of **idiom**, and they are often included in dictionaries of idioms. They share with idioms the non-literalness of their intended reference, but they differ from idioms in that their literal meaning bears a direct, though pictorial, relation to their intended reference. There is a direct parallel between 'a new broom sweeps clean' and the new person coming in and changing everything especially people's cherished ways of doing things. There is no such parallel between 'kick the bucket' and *die*, still less between 'storm in a teacup' and 'unwarranted fuss'. The essential feature of an idiom is its non-literal, metaphorical meaning. The meaning of an idiom is not the sum of the meaning of its parts, its constituent words. The meaning is idiomatic; a foreign learner has to learn the meaning of an idiom over and above the meanings of the words that make it up. With a proverb it is possible to guess the meaning by knowing the meanings of its parts and appreciating the allegorical reference.

Another characteristic of idioms is that they are fixed expressions, though this fixity is in some cases relative. An idiom like 'a storm in a teacup' is fixed. We cannot make either of the nouns plural (*'storms in a teacup' or *'a storm in teacups'), nor can we alter the sequence of the words (*'a teacup had a storm in it', etc.), nor transform it in any other way. This fixity of idioms allows us to use them as the basis for jokey comments, e.g. 'storm in an eggcup'. In the case of 'kick the bucket', however, the verb may take on different forms appropriate to the context (*kicked, has/had kicked*), but the noun may not be made plural (*'kick the buckets'), nor may the clause be made passive (*'the bucket was kicked'). With the idiom 'let the cat out of the bag' ('give away a secret') on the other hand, the passive is allowed ('the cat has been let out of the bag'), though the nouns may not be made plural (*'let the cats out of the bags'). Idioms thus differ in how fixed they are, in the sense that the extent to which idioms may be grammatically manipulated is variable.

We have cited non-literal or metaphorical meaning as a necessary characteristic of an idiom. Some lexicologists would argue that 'pure' idioms allow both a literal and a non-literal interpretation. That is to say, an idiom is by its very nature misleading: its non-literal meaning cannot be deduced from its literal one. So 'kick the bucket' could be interpreted literally, but its literal interpretation bears no relation to *die*. We have to know that in the appropriate context 'kick the bucket' means *die*. A similar example is 'let the cat out of the bag', where the literal interpretation is misleading, and you need to know the non-literal meaning to interpret it appropriately when the expression is used as an idiom. Not all fixed expressions, however, that we would call idioms have this ambiguity between literal and non-literal interpretation. 'A storm in a teacup', for example, has no literal interpretation; it is used only metaphorically, though its non-literal meaning cannot be deduced from the meanings of its constituent words.

Look now at the idioms in [9]. Which of these have both a literal and a non-literal interpretation, and which have only a non-literal meaning?

[9a] cross swords with someone
[9b] dyed in the wool
[9c] fly off the handle
[9d] out of the blue
[9e] put the wind up someone
[9f] run out of steam
[9g] spill the beans
[9h] throw someone in at the deep end
[9i] up to the ears in something
[9j] wash one's hands of something

Of the idioms listed in [9], those which could be said to still have a readily recognisable literal meaning are: a, c, g, h, j. We might add [9f] to this list, though the passing of even memories of the steam age (when an engine might run out of steam and be brought to a halt) will render the literal interpretation of this expression less and less likely. The remaining idioms (b, d, e, i) probably no longer have a possible literal meaning for most speakers of English.

With idioms, then, we cannot look at the individual words of the expression and describe the contribution that each makes to the meaning of the whole. We have to consider the meaning of the expression as a unity. This is not the case with one further kind of fixed expression: the conventional **simile** (e.g. 'as sly as a fox'). A simile is composed of a part that is interpreted literally (e.g. *sly*) and a part that is interpreted more or less non–literally (e.g. *fox*). In our culture we have conventionally imputed certain characteristics to non–human creatures (e.g. slyness to foxes), which we then draw on as comparisons (similes) for human beings. When we say of someone that they are 'as sly as a fox', we do not imply that they are a fox, but that they share a characteristic which we have culturally imputed to foxes. These comparisons are then expressed in a number of fixed phrases, so that 'as sly as' must always be completed by 'a fox'. There is no problem of interpretation with similes, as there is with idioms; and each word can be taken at its face value as far as its contribution to the meaning of the whole expression is concerned. For this reason similes are not usually entered in dictionaries.

Collocation and idiom in the dictionary

If dictionaries contain collocational information then it is derived from the accumulated informal knowledge of lexicographers, since, as we have noted, objective statistical information is not available. Dictionaries vary both in whether and in how much collocational information they give. For example, the *Collins English Dictionary* (*CED*) contains as part of its definition of *rancid* '(of butter, bacon, etc.)', whereas the *Longman Concise English Dictionary* (*LCED*) contains no such indication of collocational restriction. Sometimes collocation will be indicated not by an explicit note, as in the case of *rancid* in *CED*, but by the division of senses in the entry for a lexeme. For example, one of the senses of *hiss* in *CED* is 'such a sound uttered as an exclamation of derision, contempt, etc., esp. by an audience or crowd', which indicates one of the groups of subject nouns associated with *hiss*. But this is a rather haphazard way of indicating collocation; and generally collocational meaning does not yet find a significant place in dictionaries.

Now look up the following words in your dictionary. Is there any indication of the typical collocations of these words?

[10] earn   ebb   edit   elastic   elect

All the lexemes in [10] have some indication of collocational compatibility in *CED*, and most of them, though to a lesser extent, in *LCED*. In *CED* *earn* is defined in its first sense as: 'to gain or be paid (money or other payment) in return for work or service', indicating that the range of objects of *earn* is restricted to money or some kind of payment. The first definition for *ebb* in *CED* reads: '(of tide water) to flow back or recede', indicating the collocational range of the subject. *Edit* has a number of senses; the first one in *CED* reads: 'to prepare (text) for publication . .', indicating the range of objects of *edit*, i.e. text. The second sense reads: 'to be in charge of (a publication, esp. a periodical)', again indicating the range of objects. In the case of *elastic* one sense in *CED* (the first) reads: '(of a body or material) capable of returning to its original shape after compression, expansion, stretching or other deformation'. And another sense (the fifth) reads: '(of gases) capable of expanding spontaneously', indicating in both cases the range of nouns that *elastic* may modify in each of its senses. Finally, the first sense of *elect* is defined as: 'to choose (someone) to be (a representative or public official)', specifying the range of both the object and the object complement of *elect*. Dictionaries do thus contain collocational information for some of their entries, but it is not specified in any systematic fashion.

We have noted already (Chapter 1, p. 14) that general dictionaries contain idioms, under the headword of one or more of the principal lexemes in the idiom. 'Kick the bucket', for example, is found under *kick* in both *CED* and *LCED*, and also under *bucket* in *CED*. Idioms, then, are treated rather like some derived words (see Chapter 3, p. 43), as run-on entries, even though the meaning of the idiom may bear little or no relation to the meaning of the headword. Besides being included in general dictionaries, idioms also have dictionaries devoted to them alone, e.g.

the *Longman Dictionary of English Idioms*. Proverbs are not usually included in general dictionaries, though they may find a place in an idiom dictionary.

Exercises

1. Say whether each of the following expressions is a proverb, an idiom, a simile, or a mere collocation:
   (a) a fine kettle of fish
   (b) as innocent as a dove
   (c) have one's hair frizzed
   (d) hit the nail on the head
   (e) It's an ill wind that blows nobody any good.
   (f) lay siege to
   (g) make ends meet
   (h) rugged life/manners
   (i) The proof of the pudding is in the eating.
2. What kinds of objects typically collocate with each of the following verbs?
   accuse   betray   put on   repair   sing   utter
3. What do the following idioms mean?
   (a) call a spade a spade
   (b) call it a day
   (c) draw a blank
   (d) go the whole hog
   (e) in a hole
   (f) lock, stock and barrel
   (g) mind one's p's and q's
   (h) see eye to eye with
   (i) (dressed) up to the nines
4. Examine the entry for *drive* (verb) in your dictionary. What information does it contain on collocations and idioms?

CHAPTER 8

Why Dictionaries?

Most households probably own a dictionary of some kind or other, though we shall leave the discussion of what purposes it may be used for until Chapter 13, p. 192. However, before mass readership developed with the wider circulation of books following the invention of printing in the fifteenth century, there was no need or demand for dictionaries. The first book akin to the modern (monolingual) English dictionary (there were earlier bilingual dictionaries) appeared in the early seventeenth century, more than a century after printing was invented and multiple copies of books became more easily available. And it was really not until the middle of the eighteenth century, with the publication of Samuel Johnson's dictionary, that the pattern was set for dictionaries as we know them today.

Before Johnson

The earliest 'lists of words' that might be said to constitute the beginnings of English lexicography were the glossaries of Anglo-Saxon priests and schoolmen, compiled to enable those whose competence in Latin was lacking to read Latin manuscripts. Latin was the language of the church and of learning generally, and access to learning required the mastery of Latin. Probably at first, English glosses of Latin words were provided between the lines of Latin manuscripts (interlinear glosses). Then the glosses from several manuscripts were combined into a glossary, a list of difficult Latin words with their English equivalents. The more extensive a glossary became the more difficult it would be to find a particular item readily. It was this need that gave rise to alphabetisation, the arrangement of entries in the list by alphabetical order.

At first, alphabetical ordering was by the first letter of words only. So, all the words beginning with 'A' would be together in the glossary, but within the set of *A*-words the order would be random. This alphabetisation would presumably have been carried out by scribes making a new copy of a glossary. An early eighth-century glossary is alphabetically ordered to include the second letter of the words in addition to the first, and a tenth-century glossary has alphabetical ordering as far as the third letter. It might be noted at this point that alphabetical ordering has no lexicographical significance, despite the term 'dictionary order' as a synonym for 'alphabetical order'. The alphabetical ordering of word lists has a pragmatic function: to enable the user to gain ready access to the item that is being looked up.

The glossaries were essentially lists of Latin words with English glosses. It is not until the Renaissance in the fifteenth and sixteenth centuries that the reverse is found, i.e. English-Latin word lists. Again the aim is to enable learners to master Latin in order to participate in the revival of classical learning and literature. The *Promptorium parvulorum* ('children's store-room') of around 1440 has about twelve thousand entries in alphabetical order, though under each letter the 'verba' (verbs) are listed separately from the 'nomina' (nouns and words belonging to other word-classes). One of the first printed English-Latin lexicons, John Withals' *Shorte Dictionarie for Younge Begynners* of 1553, has a thematic arrangement of its words (see Chapter 14). We might note the use of the term 'dictionary' in the title of this book; it was a sixteenth-century borrowing from the Latin *dictionarium* ('collection of words'). It is in these English-Latin dictionaries of the Renaissance that we should perhaps recognise the beginnings of the lexicography of English.

The next step towards a monolingual English dictionary (i.e. one in which English words would be defined by other English words rather than by equivalents in another language) is to be seen in the bilingual dictionaries of vernacular languages in the late Renaissance.

Eventually the attention paid to the languages of classical antiquity was extended to the living languages of Europe and resulted in bilingual or even polyglot dictionaries, such as: John Florio's *A Worlde of Wordes* of 1598, an Italian-English dictionary; Randle Cotgrave's *A Dictionarie of the French and English Tongues* of 1611; and John Kinshieu's polyglot *Ductor in Linguas . . . The Guide into the Tongues* of 1617.

The first monolingual English dictionary is reckoned to be Robert Cawdrey's *A Table Alphabeticall*, published in 1604. Cawdrey's dictionary contains almost three thousand items, with short definitions of each of them. The items included are what Cawdrey calls 'hard usuall English wordes, borrowed from the Hebrew, Greeke, Latine, or French &c'. And he has collected them together 'for the benefit & helpe of Ladies, Gentlewomen, or any other unskilfull persons', so that they might understand these hard English words 'which they shall heare or read in Scriptures, Sermons, or elsewhere'. A wider reading public had been developing in Elizabethan England, and to enable its members to understand what they read, dictionaries of 'hard words' were compiled to explain mainly words borrowed as a result of Renaissance influence. Other hard-word dictionaries followed: John Bulloker's *An English Expositor* in 1616, Henry Cockeram's *The English Dictionarie* in 1623, and Thomas Blount's *Glossographia* in 1656.

The beginnings of English lexicography then established a tradition of hard-word dictionaries. If only 'hard' words were included in modern dictionaries – and they are probably the items most often looked up – their size would be greatly reduced. We will take at random two pages from the *Longman Concise English Dictionary* (*LCED*) with the headwords from *gratify* to *green*, amounting to eighty-three lexemes in all. Of these we can identify at most twenty-three that might qualify as 'hard' words (i.e. words that an average reader might not be familiar with or might need to look up in a dictionary), that is, approximately twenty eight per cent of the items. If those proportions were to be repeated throughout the *LCED*, it would contain fewer than five hundred pages instead of in excess of sixteen hundred. It is an interesting question whether there might be a justification for hard-word dictionaries today. But we need first to examine the tradition in which modern general-purpose dictionaries are to be found.

The hard-words tradition of English lexicography lasted about a century. Then in 1702 there appeared *A New English Dictionary* by one 'J.K.', who is widely though not conclusively presumed to be John Kersey, who in 1708 published a *Dictionarium Anglo-Britannicum* under his full name. The *New English Dictionary* declares itself 'a Compleat collection of the most proper and significant words, commonly used in the language; with a short and clear Exposition of Difficult Words and Terms of Art'. The most significant word in this quotation is 'compleat'. Here we begin the tradition of comprehensiveness that is with us still. J.K. aims to include 'all the most proper and significant words', but they are to be restricted to 'such English Words as are genuine, and used by Persons of clear Judgment and good Style'. The aim is that a 'plain Country-man' might find the common English words. The principle of comprehensiveness is connected with providing information and advice for the linguistically gauche or insecure.

A new departure which reinforced the tradition of comprehensiveness was made by Nathaniel Bailey with his *An Universal Etymological English Dictionary*, published in 1721. As his title suggests, Bailey paid considerable attention in his dictionary to the origins of words. Clearly, if lexicographical description was to include information on the etymologies of words, then there was every reason for all the words in the language to be included in a dictionary, if only for the sake of their etymology. Bailey's dictionary contained some forty thousand words. This was increased to some forty-eight thousand in his *Dictionnarium Britannicum* of 1730. The tradition of comprehensiveness had been established, a tradition in which Samuel Johnson found himself as he compiled his dictionary in the mid-eighteenth century.

Samuel Johnson

The task of advising ignorant users of the language was taken up with considerable vigour by Samuel Johnson. In

1747 he published *The Plan of a Dictionary of the English Language*, addressed to the Earl of Chesterfield, whose patronage he hoped – in vain however – to win for the project. For the next seven years Johnson toiled at his *Dictionary of the English Language*, whose first edition was published in 1755 in two volumes. A comparison of the *Plan* – what Johnson intended to do in his dictionary – and the Preface to the *Dictionary* – what he actually did – is instructive.

Johnson states his intentions in the *Plan* as follows:

[1] The chief intent of it is to preserve the purity and ascertain the meaning of our English idiom . . . one great end of this undertaking is to fix the English language . . . This . . . is my idea of an English Dictionary, a dictionary by which the pronunciation of our language may be fixed, and its attainment facilitated, by which its purity may be preserved, its use ascertained, and its duration lengthened. And though . . . to correct the language of nations by books of grammar, and amend their manners by discourses of morality, may be tasks equally difficult; yet it is unavoidable to wish, it is natural to hope . . . that it may contribute to the preservation of antient, and the improvement of modern writers . . . and awaken to the care of purer diction, some men of genius, whose attention to argument makes them negligent of style.

Additionally he hoped to fix the spelling of words, which, although they had stabilised considerably, still contained inconsistencies. The dictionary would be authoritative because it would be based on citations from 'writers of the first reputation'. Here is Johnson's most notable innovation: the use of citations from English literature in order both to illustrate the usage of words and to build up the definitions in the first place. Although he used Bailey's dictionary as the basis for his work, it was supplemented, augmented and revised by Johnson's examples from the 'best' literature.

Although in the Preface to the *Dictionary* he is still about the task of preservation – 'tongues, like governments, have a natural tendency to degeneration; we have long preserved our constitution, let us make some struggles for our language' – Johnson is less sanguine about the possibility of fixing the language. If you compare the following quotation from the Preface to the *Dictionary* with that from the *Plan* in [1], some differences of attitude can be perceived.

[2] Those who have been persuaded to think well of my design, will require that it should fix our language, and put a stop to those alterations which time and chance have hitherto suffered to make in it without opposition. With this consequence I will confess that I flattered myself for a while; but now begin to fear that I have indulged expectation which neither reason nor experience can justify. When we see men grow old and die at a certain time one after another, from century to century, we laugh at the elixir that promises to prolong life to a thousand years; and with equal justice may the lexicographer be derided, who being able to produce no example of a nation that has preserved their words and phrases from mutability; shall imagine that his dictionary can embalm his language, and secure it from corruption and decay, that it is in his power to change sublunary nature, and clear the world at once from folly, vanity, and affectation.

Johnson acknowledges in the Preface that he had for a time cherished the illusion that he might be able to fix and preserve the language by means of his dictionary. This illusion is expressed in the *Plan*, though he acknowledges it to be an all but impossible task: 'Though art may sometimes prolong their [words'] duration, it will rarely give them perpetuity, and their changes will be almost always informing us, that language is the work of man, of a being from whom permanence and stability cannot be derived'. Eight years later, having worked relentlessly on the *Dictionary*, he can recognise it clearly as an illusion. But though change, which he equates with 'corruption and decay', cannot be stopped, it is worthwhile to make some attempt to counter the decay: 'It remains that we retard what we cannot repel, that we palliate what we cannot cure'. The *Dictionary* was to play its part in that attempt. It would provide an authority that could be referred to, a standard by which usage could be judged. Johnson contributed to the establishment of a further lexicographic tradition that persists to today: the dictionary as an authority on standards of usage.

Samuel Johnson's concerns must be seen in the light of eighteenth-century attitudes to language generally. There was a strong movement to form an Academy on the model of the Italian 'Accademia della Crusca', founded in 1582, and the French 'Académie Française', founded in 1635, which were both devoted to 'purifying' their respective languages. Daniel Defoe argued for an English Academy; so did Joseph Addison, one of the founders of the *Spectator*. But the strongest voice in favour of an Academy was that of Jonathan Swift, who in his *Proposal for Correcting, Improving, and Ascertaining the English Tongue* championed the formation of an Academy which would be charged with the task of devising a method 'for Ascertaining and Fixing our Language for ever, after such alterations are made in it as shall be thought requisite'. Johnson was against the formation of an Academy; he thought that it would be difficult to set up and that its authority would not be respected by the public at large. Good linguistic manners would not be derived from the authoritarian pronouncements of an Academy but from the precedents set by good writers from the past. And it was on those precedents that his *Dictionary* had been based, as he tells us in the Preface:

> [3] . . . I have studiously endeavoured to collect examples and authorities from the writers before the restoration, whose works I regard as 'the wells of English undefiled', as the pure sources of genuine diction.

Meanwhile, on the other side of the Atlantic Ocean, another community was continually expanding, whose language was also English. Here a debate developed between those who wished to derive standards of usage from the linguistic practice of the mother-country and those who wished for the English language in America to develop in its own way and derive its own standards and authorities. Chief protagonist of the second view in the late eighteenth and early nineteenth centuries was Noah Webster, who not only attempted to introduce a wide ranging spelling reform but also set about compiling a dictionary which would do for American English what Johnson's had done for the English of Britain. If Johnson had taken the best British writers as his authorities, Webster would take the best American writers. And Webster's *American Dictionary of the English Language*, published in two volumes in 1828, is very much in the Johnson tradition, as Webster expresses it in the Preface to his Dictionary:

> [4] It has been my aim in this work, now offered to my fellow citizens, to ascertain the true principles of the language, in its orthography and structure; to purify it from some palpable errors, and reduce the number of its anomalies, thus giving it more regularity and consistency in its forms, both of words and sentences; and in this manner, to furnish a standard of our vernacular tongue, which we shall not be ashamed to bequeath to three hundred millions of people, who are destined to occupy, and I hope, adorn the vast territory within our jurisdiction.

The New English Dictionary

Following Nathaniel Bailey, Samuel Johnson had included etymologies in his *Dictionary*. In the *Plan*, he had signalled the intention of providing even more of the history of words by arranging 'the quotations . . . according to the ages of their authors', showing when a word was first introduced and for an obsolescent word when it was last attested:

> [5] . . . the reader will be informed of the gradual changes of the language, and have before his eyes the rise of some words, and the fall of others.

But Johnson was doubtful whether such detail could be provided. This would indeed have to wait for the great nineteenth-century enterprise which culminated in the *Oxford English Dictionary* in twelve volumes.

The nineteenth century was above all the century interested in history and origins, in language as in other disciplines. Johnson's intention of providing a history of words by means of chronologically ordered citations was carried a step further by Charles Richardson, who published his *New Dictionary of the English Language* in 1836/7. Its title page proclaimed it to be 'Illustrated by Quotations from the Best Authors'. In it definitions are all but eclipsed by the chronological sequences of quotations from the Middle English period onwards, illustrating the changes in meaning that words had undergone. But, as Johnson himself had realised, the amount of data needed to compose an accurate and comprehensive history of all the words of English was beyond the ability of one person to amass.

The initiative to compile a thoroughgoing historical dictionary came from the Philological Society, which in 1857 appointed a committee 'to collect unregistered words in English', with the intention of producing a supplement to existing dictionaries such as Richardson's. Later that same year one of the members of the committee, Richard Chenevix Trench, Dean of Westminster, read a paper to the Society, entitled 'On Some Deficiencies in our English Dictionaries'. In it he articulated clearly the 'historical principle' as the only sound basis for lexicography and outlined what he thought a dictionary should contain. To the Philological Society the idea of a supplement now seemed rather inadequate, and early in 1858 it resolved to commission the preparation of a *New Dictionary of the English Language*. During the next twenty years, under the editorships of Herbert Coleridge and F. J. Furnivall, various kinds of preparatory work were undertaken – compiling lists of literary works to be read for quotations, deciding on the organisation of entries, collecting and arranging quotations – but it was all done on a part-time basis, with subeditors in various parts of the country, and by 1878 the project had got into the doldrums.

In 1878 two new factors brought the *New English Dictionary* to life again. The first of these was the appointment of James A.H. Murray to the editorship. The second was the negotiation of a contract with the Oxford University Press to publish the dictionary. Murray was to be allowed ten years in which to complete the work, even though he was to continue his job as a schoolmaster. Before the dictionary was complete, however, Murray had become full-time editor and moved to Oxford, and three further editors were appointed: Henry Bradley, W. A. Craigie, and C. T. Onions. And although publication of the dictionary began in 1885 it was not completed until 1928, by which time Murray and Bradley had died. From 1895 the designation *The Oxford English Dictionary (OED)* began to appear on the title page, and this name gradually replaced the original *New English Dictionary on Historical Principles*.

When Murray gathered together the material that had been amassed before his appointment to the editorship, he realised that there was still much to be done by way of collecting quotations before the dictionary entries could be written. A new appeal for voluntary readers was launched and a new set of directions was issued:

[6] Make a quotation for *every* word that strikes you as rare, obsolete, old-fashioned, new, peculiar, or used in a peculiar way. Take special note of passages which show or imply that a word is either new and tentative, or needing explanation as obsolete or archaic, and which thus help to fix the date of its introduction or disuse. Make *as many* quotations *as you can* for ordinary words, especially when they are used significantly, and tend by the context to explain or suggest their own meaning.

In the event some five million excerpts were sent in to the editorial office and they formed the basis on which the dictionary entries were composed. Some one million eight hundred thousand of them are actually cited in the published dictionary. The tradition of collecting excerpts containing words or senses of words so far unrecorded has been continued by the Oxford journal *Notes and Queries*, which regularly contains articles on new lexicographic finds.

The Preface to the *OED* states the aim of the dictionary as: 'to present in alphabetical series the words that have formed the English vocabulary from the time of the earliest records down to the present day, with all the relevant facts concerning their form, sense-history, pronunciation, and etymology'. It is acknowledged that the notion of 'the English vocabulary' is indeterminate and that the lexicographer must make choices to include and exclude, which may be arbitrary. The *OED* aims to include 'all the "Common Words" of literature and conversation, and such of the scientific, technical, slang, dialectal, and foreign words as are passing into common use, and approach the position or standing of 'common words'. This aim is of course combined with the historical perspective, so that words that were once in common use but at some time ceased to be so are marked as obsolete. In fact the *OED* excludes any word which was obsolete by 1150. For those included, their history is charted from their first observed appearance in common use to their last observed appearance: those in current use would have citations from the second half of the nineteenth century.

Clearly, such a dictionary is in constant need of updating, and when the *OED* was reissued in 1933 in its now familiar twelve-volume format, a Supplement volume was also published, containing new words and meanings as well as additions and amendments to the original volumes. Since then the English vocabulary has expanded enormously, especially in the scientific and technical areas, but also as a result of the use of English internationally. A *Supplement to the Oxford English Dictionary* has been published in four volumes, edited by R. W. Burchfield, to replace the 1933 Supplement. Volume One, containing the letters 'A' to 'G' appeared in 1972, followed by Volume Two ('H'–'N') in 1976, Volume Three ('O'–'Scz') in 1982, and Volume Four ('Se'–'Z') in 1986. Now the files of the *OED* are being computerised, and regular updating will become a more manageable process.

Many people look to the *OED*, because of its undoubted prestige as the most extensive description of the vocabulary of English, as the ultimate authority in matters of current usage. But two points need to be taken account of. The first is that the *OED* is primarily an historical dictionary, tracing the development of the forms and meanings of the words that go to make up the vocabulary of English. The second is that in its original twelve-volume form it represents the vocabulary current at the end of the nineteenth century. The British public is perhaps less demanding that a dictionary should be authoritative than is the American public, and this is illustrated nowhere better than in the reception accorded to the *Webster's Third New International Dictionary*.

The rights to the Webster dictionaries were purchased after his death by G. & C. Merriam Company. In 1890 the term 'international' replaced 'American' in the title, and the first of the 'New International' dictionaries was published in 1909. The *Second* followed in 1934, and the *Third* in 1961, edited by Philip B. Gove. *Webster's Third* continued in the inductive tradition of lexicography, begun by Samuel Johnson, taken up by Noah Webster, and supremely evident in the *OED*, namely the description of words on the basis of examples of usage. 'The definitions in this edition', Gove states in the Preface, 'are based chiefly on examples of usage collected since publication of the preceding edition.' Excerption by editorial staff added four-and-a-half million quotations to the more than one-and-a-half million already on file from previous editions. Together with citations from other dictionary sources (including the *OED*) and concordances, Gove estimates that over ten million citations form the basis for the definitions in *Webster's Third*. This use of textual evidence, as well as the absence of prescription and a generally synchronic rather than diachronic (or historical) orientation, resulted from the conscious influence of modern descriptive linguistics on the editors of *Webster's Third*.

According to Gove, *Webster's Third* is intended 'as a prime linguistic aid to interpreting the culture and civilisation of today, as the first edition served the America of 1828'. It is 'the record of this language [English] as it is written and spoken'. There is no hint of the dictionary's function as prescriptive authority in these pronouncements. Authority there is, though, but it arises from the dictionary's accuracy:

> [7] . . . the editors of this new edition have held steadfastly to the three cardinal virtues of dictionary making: accuracy, clearness, and comprehensiveness. Whenever these qualities are at odds with each other, accuracy is put first and foremost, for without accuracy there could be no appeal to WEBSTER'S THIRD NEW INTERNATIONAL as an authority. Accuracy in addition to requiring freedom from error and conformity to truth requires a dictionary to state meanings in which words are in fact used, not to give editorial opinion on what their meanings should be.

It was at this point that many reviewers and members of the American public felt that their Dictionary had abrogated its public duty to tell them how they ought to use words. Criticism focused in large part on what the

Dictionary included – part of its recording function. There appeared to be a widespread belief that if a word or a meaning had no place in educated or 'refined' language, then it should have no place in the Dictionary either. Repeatedly, the inclusion of *ain't* was cited as a particularly glaring example of the Dictionary's indiscretion. In fact, *ain't* had appeared in the *Second New International*, though with a very brief entry and with the usage labels 'Dial.' (dialect) and 'Illit.' (illiterate). These are of course terms of lexicographic description, though they were interpreted by dictionary users as evaluative terms. The *Third* gives *ain't* much fuller treatment and includes the descriptive comment:

[8] though disapproved by many and more common in less educated speech, used orally in most parts of the U.S. by many cultivated speakers, esp. in the phrase *ain't I*.

Reviewers felt that the Dictionary had made *ain't* respectable; but they still disapproved of *ain't*, and the Dictionary should therefore have condemned it, as they believed the *Second* had done.

A further focus of criticism was the quotations. The editors had not been sufficiently selective in their sources of quotations. No longer were they accepted just from the 'best authors', but many of the contemporary quotations came from ephemeral publications or from authors who were nonentities. In the Preface, Gove speaks of 'a systematic reading of books, magazines, newspapers, pamphlets, catalogs, and learned journals'. For many this was a clear indication that the Dictionary was allowing standards to slip and could no longer be trusted. *Webster's Third* offered them a record of current English usage; what they wanted was an authoritative standard which told users what the best usage should be.

These criticisms bring us back to the question posed in the title of this chapter: why dictionaries? If we go back to Cawdrey's *Table Alphabeticall* and other early seventeenth-century dictionaries, then we can identify the aim as one of supplying unsophisticated readers with explanations of unfamiliar, mostly borrowed words. The same aim, though more broadly conceived, is reflected in the intention of *Webster's Third* to be 'a prime linguistic aid to interpreting the culture and civilization of today'. Dictionaries have the role of handbook, of reference work, so that speakers, or more especially readers, of the language can find help in understanding words that they come across for the first time and whose meanings need explanation. Such a need could arguably be best met by a publication containing a judicious selection of the 'hard' words, in the tradition of Cawdrey.

If we go back to Nathaniel Bailey's *An Universal Etymological English Dictionary* and Samuel Johnson's *Dictionary of the English Language* we find another, altogether different aim. In these cases the dictionary is intended to provide a record of the history of English words, particularly a history of the derivation of the forms of words (etymology), but for Johnson a history of the development of the meanings of words (sense-history) as well, though he did not have the resources to achieve this in any systematic fashion. It was only the immense undertaking that the *Oxford English Dictionary* became that could produce what Johnson foresaw as the semantic complement to etymology. The dictionary was now seen as having a recording function, and particularly – in the case of the *Oxford English Dictionary* supremely – the recording of the history of words. Consequently, dictionaries expanded enormously, and comprehensiveness, if not completeness, of inclusion became the order of the day.

Webster's Third also claims to be 'a record of [the] language as it is written and spoken'. But the record is no longer historical. The entries in *Webster's Third* contain etymologies, but they do not give sense-histories. The record has now become the record of the contemporary language: the state of the current vocabulary, or at least the cross-section that a particular dictionary is big enough to cope with. Modern dictionaries thus have a double function: they represent on the one hand a piece of linguistic description, an account of the landscape of English words; on the other hand they are purveyed as manuals for users of the language to refer to particularly when they need help in understanding what they read.

These two functions are, however, complexly interwoven with a third. Johnson saw his dictionary as setting a standard for linguistic usage; the standard would derive from the practice of the 'best writers', whom he had excerpted for illustrative quotations. Future lexicographers may deny that they see their dictionaries as setting the same authoritative standard that Johnson intended his to be: for the dictionary-buying public, 'the dictionary' is not just a reference manual or a record of the vocabulary; it provides an authority on how the language should be used, and they have recourse to it as an arbiter in disputes about linguistic usage. The dictionary has an imputed function as the authority on the language.

Exercises

1. Read Johnson's *The Plan of a Dictionary of the English Language* and/or his Preface to the *Dictionary*. (The *Plan* is reprinted in M. Wilson (ed.), 1957. The Preface is also reprinted there and is found in most anthologies of Johnson's prose.) What different aspects of the description of a word (e.g. spelling, pronunciation) would he deal with in his lexicographical description? If you have access to a copy of Johnson's *Dictionary*, find out how far these aspects are actually dealt with.

2. Read the 'General Explanations' in the front-matter of the *Oxford English Dictionary*. (You may also for interest like to read the 'Historical Introduction'.)
   (a) What are the differences between 'main words', 'subordinate words' and 'combinations'?
   (b) What aspects of the description of words does the *OED* include?
   (c) What are 'Naturals', 'Denizens', 'Aliens' and 'Casuals'?

3. Read the 'Preface' to *Webster's Third New International Dictionary*. How does this dictionary deal with the description of (a) pronunciation and (b) etymology?

4. If you have access to Johnson's *Dictionary*, *OED* and *Webster's Third*, look up a particular word (e.g. *noise*) in all three and find out how far the three entries reflect the different aims of the dictionaries.

5. If you were given the task of composing a dictionary, what elements would you give preference to, and why? Would you introduce any new features? If so, which and why?

CHAPTER 9

How to Define a Word

Dictionaries are popularly conceived as reference works in which we look up the meaning of words. Giving meanings is seen as the central function of dictionaries. And dictionary definitions are accounts of meaning, the attempt to express the meaning of each word distinctively. However, as we have seen in Chapters 5 and 7, the meaning of a lexeme involves not just what it denotes intrinsically, but also its relations with other words of similar or opposite meaning in the same lexical field or semantic domain, as well as its relations with words that regularly cooccur with it in collocations. Although dictionary definitions do on occasions take account of both these aspects of meaning, the alphabetical arrangement of the dictionary restricts the extent to which lexical field relations can be fully invoked, and information on collocations is at a relatively elementary stage for it to be of much benefit to lexicographers.

Dictionary definitions must therefore be viewed as provisional, as representing the potential meaning of a word, waiting for actualisation in a context. There is, in fact, a problem with the term 'definition'. If we ask for a definition of say a technical term in one of the natural sciences (e.g. 'chromosome' in genetics) we expect an exact characterisation of the term – of the way it is used in the descriptive framework of that field of study – such as we might find in an encyclopaedia. But ordinary language is not like the languages of the sciences. Its words are not normally used with the same precision and accuracy that their words are. The meanings of ordinary words are to varying extents indeterminate and fuzzy. Consequently, a dictionary definition is not to be thought of as giving a complete characterisation or a complete semantic analysis of a lexeme.

There is a further problem with dictionary definitions that needs to be mentioned here. It is a problem that is inherent to the whole enterprise of linguistic description. It is that in writing the definition of a word we have to use other words in order to do so. The phenomenon to be described and the means of description are both language. In other sciences the means of description is language, but what is described is chemical matter, or geographical features, or social behaviour, and so on. Dictionary definitions are, then, forms of paraphrase: putting a word into other words. In bilingual dictionaries the 'other words' are translation equivalents, but in monolingual dictionaries, which are our chief concern, the 'other words' are some kind of paraphrase in the same language. In this chapter we are going to examine what is involved in defining lexemes.

Establishing separate meanings

Many lexemes, especially those in common use, have multiple meanings: they are polysemous. Polysemy, however, has to be distinguished from homonymy (see Chapter 1, p. 5). Homonyms (or homographs) need separate entries in the dictionary. For example, *ear* (= 'organ of hearing') and *ear* (= 'part of a cereal plant, e.g. corn') are homonyms; they are not related in meaning and they have different etymologies; so they constitute two separate headwords. On the other hand, *foot* (= 'projection of the leg') and *foot* (= 'bottom part of page, mountain, etc.') are polysemes; the second represents an extension of the meaning of the first; and so they are entered as variant meanings (senses) of the same headword. However, the distinction between homonymy and polysemy is not always clearcut: there are instances where it is not easy to decide whether we have a case of homonymy or of polysemy.

For example, the word *pope*, besides referring to the holder of a particular ecclesiastical office, is also used to refer to a kind of small freshwater fish, also called *ruff* (or *ruffe*). This is presumably a case of extension of meaning, from the pontiff to the fish, the borrowing of a term from the ecclesiastic realm into the piscatorial field on the basis of some supposed resemblance. Do we then treat it as a case of polysemy on the basis of common etymology, or do we treat it as homonymy on the basis of a clear difference in denotational meaning? If it is treated as polysemy, there will be a single headword *pope*: if it is treated as homonymy there will be two headwords. Look this item up in your dictionary to see how it is treated.

In the *Longman Concise English Dictionary* (*LCED*), *pope* is treated as a case of polysemy, but in the *Collins English Dictionary* (*CED*) it is treated as homonymy, with two headwords. Usually the criterion for deciding between polysemy, or a single headword, and homonymy, or more than one headword, is etymology. If the meanings of an item (orthographic word) can be shown to be derived from a common origin, then this is treated as polysemy, even if resultant meanings diverge considerably. On this basis, for example, *nail* (= 'horny growth on fingers and toes') and *nail* (= 'pointed fastening device knocked in by a hammer') are treated as different senses of the same headword. On the same basis, however, as we have seen, *ear* (= 'organ of hearing') and *ear* (= 'part of cereal plant') constitute different headwords. With *pope*, where some doubt exists about the relationship between the meanings, there is room for divergence of lexicographic practice.

Decisions about polysemy presuppose that the different meanings or senses of a lexeme have been identified, and we may ask how lexicographers go about identifying the different senses of lexemes. If an inductive method is followed, as in the case of the *Oxford English Dictionary* (*OED*) and *Webster's Third New International Dictionary* (see Chapter 8, p. 120 and p. 122), then the lexicographer begins by assembling all the quotations collected for a particular lexeme. These are then sorted so that quotations which clearly refer to the same meaning are put together, and in this way a division into senses should emerge. Much will depend, however, on the experience, informed judgements and linguistic intuitions of the lexicographer. And so it is not to be expected that every dictionary (of similar size) will subdivide the meaning of a lexeme in the same way or into the same number of senses, especially in the case of commonly used words (e.g. *round, set*) with large numbers of identifiable senses.

Dictionaries smaller than *OED* or *Webster's Third* are often derivative dictionaries. This means that they are compiled on the basis of existing dictionaries, using the already digested information of usually larger works. This apparent plagiarism has a long pedigree in lexicography, going back to the beginnings of monolingual dictionaries in the seventeenth century. And it has frequently been practised within families of dictionaries, where a desk-dictionary may be derived from a larger one and may in turn give rise to a concise or pocket edition. Its disadvantage is that, pursued uncritically, it can perpetuate errors from one generation of dictionaries to the next. In derivative dictionaries, the division of meaning into senses will be based on previous work, modified by the judicious selection and careful scrutiny of the lexicographer.

To illustrate the differing treatments of the same lexeme, look up the following words in two up-to-date dictionaries of a similar size from different publishers, and examine how many senses each has been divided into and the extent of overlap of the senses identified.

[1] chauvinism   holocaust   venerable

LCED identifies two senses of *chauvinism*: 'excessive or blind patriotism' and 'undue attachment to one's group, cause or place'. *CED* identifies three senses, with the first identical to the first sense in *LCED*: the second *LCED* sense has been divided into two in *CED*: 'enthusiastic devotion to a cause' is separated from 'smug irrational belief in the superiority of one's own race, party, sex, etc.'. In the case of *holocaust*, both dictionaries identify the senses 'great destruction or loss of life' and 'burnt offering', in that order in *CED*, but reversed in *LCED*. *LCED* adds a third sense referring to the 'genocidal persecution of European Jewry . . . during WWII'. For *venerable, LCED* identifies three senses, and *CED* recognises five, largely through further subdividing *LCED* senses. The ecclesiastical title takes up one sense in *LCED* and two in *CED*, which divides the Anglican from the Roman Catholic use. The dictionaries share the sense of 'hallowed by religious or historical association'. The third sense in *LCED* has two subdivisions: **a** 'commanding respect through age, character, and attainments'; **b** 'impressive by reason of age'. The **a** subdivision corresponds to the first sense in *CED* ('(esp. of a person) worthy of reverence on account of great age, religious associations, character, position, etc.'), while the **b** subdivision would appear to correspond to the third sense in *CED* ('ancient'). Clearly there is a fair degree of scope for variously dividing the meanings of lexemes into senses, and it must be seen as part of the art, or craft, of lexicography (see Chapter 15).

We will now turn to the question of how the senses of a lexeme are ordered in a dictionary entry. We noticed that *LCED* and *CED* differed in their ordering of the senses of *holocaust* and *venerable*. Let us take a further example: *quash*. *LCED* identifies two senses for *quash*, and the first sense is subdivided into two, as follows:

[2] **1a** to nullify (by judicial action) **b** to reject (a legal document) as invalid **2** to suppress or extinguish summarily and completely; subdue

The first sense is the legal one, and the second is the more general one. In *CED*, *quash* has three senses, because the **a** and **b** of *LCED* are given separate numbers, as follows:

[3] **1.** to subdue forcefully and completely; put down; suppress. **2.** to annul or make void (a law, decision, etc.). **3.** to reject (an indictment, writ, etc.) as invalid.

It will be noted that the second sense of *LCED* is placed first in *CED*. This is because the two dictionaries operate different policies on the ordering of senses within a dictionary entry. *CED* places first the sense 'most common in current usage' (Guide to the Use of the Dictionary, p. xv), with subsequent senses 'arranged so as to give a coherent account of the meaning of a headword', but with the senses grouped together for each word-class (part-of-speech) of which the lexeme is a member. In *LCED*, on the other hand, unrestricted senses are given before restricted ones (e.g. dialectal or archaic), but they are given in historical order, with older senses preceding newer ones. The legal sense of *quash* is older than the more general sense, though less commonly used in current English (at least in the opinion of the *CED* editors).

Division into senses will also vary according to how

the dictionary treats the membership of different word-classes by the same word, e.g. when *skin* belongs both to the noun and to the verb class. Some dictionaries (e.g. *CED*) apparently consider that an extension of meaning has taken place and therefore treat such cases as different senses of the same lexeme. Other dictionaries (e.g. *LCED*) apparently recognise a process of derivation (conversion, see Chapter 2, p. 32) and therefore treat them as cases of homonymy, with a separate headword for each word–class (*i.e. 1skin* n, *2skin* vb). The division into senses thus depends on a number of decisions for which policies need to be formulated by the editors of a dictionary, and for which lexicographers need to possess great insight and good judgement.

Methods of defining

Although considerable skill and experience is required in the writing of definitions, this is often undertaken according to conscious policy or within a recognised tradition of defining. For example, Gove states the defining policy of *Webster's Third* in the Preface as follows:

[4] The primary objective of precise, sharp defining has been met through the development of a new dictionary style based upon completely analytical one-phrase definitions throughout the book.

The essence of the method stated here is that it is analytical and that definitions are expressed in a single phrase, e.g.

[5] **organ** . . . any of several large musical instruments producing sustained tones and played by means of a keyboard: (**1**): a wind instrument consisting of sets of pipes sounding by compressed air, controlled by manual and pedal keyboards, and capable of producing a variety of musical timbres and orchestral effects – called also 'pipe organ' (**2**): REED ORGAN (**3**): an instrument in which the sound and resources of the pipe organ are approximated by means of electronic devices.
door a moveable piece of firm material or a structure supported usu. along one side and swinging on

pivots or hinges, sliding along a groove, rolling up and down, revolving as one of four leaves, or folding like an accordion by means of which an opening may be closed or kept open for passage into or out of a building, room or other covered enclosure or a car, airplane, elevator, or other vehicle.

We might question whether the attempt to be 'completely analytical', while it conforms a definition more to what is meant by 'definition' in the sciences, may be totally appropriate to a dictionary, especially in the case of common words like *door*, which most dictionary users would probably never have the need to look up in any case. The restriction to one phrase can lead to rather complexly constructed definitions such as that in [5].

An alternative tradition of defining would attempt to 'typify' the meaning of a lexeme, rather than be analytical; though some dictionaries mix both approaches to definition. Look up the definitions for *lynx* and *lyre* in your dictionary. Do they have a comprehensive analytical definition or is the attempt rather to say what is typical about the meaning?

Consider the definitions of these lexemes in *LCED*:

[6] **lynx** . . . any of various wild-cats with relatively long legs, a short stubby tail, mottled coat, and often tufted ears

[7] **lyre** . . . a stringed instrument of the harp family used by the ancient Greeks esp to accompany song and recitation

The definition for *lynx* is considerably more analytical than that for *lyre*, though both are analytical in the sense that each assigns the lexeme to a class of items of which the lexeme is a member, wild-cats in the case of *lynx*, and stringed instruments of the harp family for *lyre*. But that is as far as the definition for *lyre* goes; it is completed by an indication of its typical function. The definition of *lynx* provides a number of additional analytical facts, concerning the legs, tail, coat and ears of the lynx. From these definitions we could build up a more accurate and detailed picture of the lynx in our minds than of the lyre. The definition for *lyre* tells us rather what is typical about this stringed instrument: its use by the Greeks for accompaniment.

A number of approaches to defining would want to emphasise that definitions should be substitutable. This means that a dictionary definition should be able to replace the lexeme in context with minimum loss of meaning and leaving the syntax as far as possible intact. Under this policy definitions have the nature of synonymic paraphrases. For example, in the sentence at [8] below we could substitute the *LCED* definition for *leapfrog* as in [9]:

[8] The children have been playing leapfrog all afternoon.

[9] The children have been playing all afternoon a game in which one player bends down and another leaps over him/her.

But this would work successfully for only a small proportion of the definitions in *LCED*. Such a policy for writing definitions, while it may be of some help to the dictionary user, places considerable constraints on the lexicographer; and there are probably a number of words that could just not be defined in this way, e.g. prepositions and determiners. Another defining policy that has been advocated, though not often followed, is to let the examples speak for themselves, with a minimum of paraphrase. The principal disadvantage of this policy is that it is highly uneconomical on space. Most dictionaries use examples as a supplement to definitions (see Chapter 10) rather than as a defining device in themselves.

Considerations of space are a constant concern of lexicographers, from the big *OED* down to the smallest pocket-size dictionary. This means that definitions need to be written as economically and concisely as possible. Clearly this requirement is more easily fulfilled under some defining policies than under others. A policy of complete analyticity, such as that pursued by *Webster's Third*, and illustrated by the definition at [5], will find the principle of economy difficult to observe. Another principle, which may conflict with that of economy, is that lexemes should be defined by words simpler than themselves. Not only may this be an uneconomical principle, it may be impossible to observe, especially where a lexeme is already 'simple'. To illustrate this point, look up the definitions in your dictionary for *embouchure* and *flower*.

It is arguable that you would be more likely to want to consult your dictionary for *embouchure* than for *flower* in any case. You have probably found that the definition of *embouchure* has been written in words simpler than the lexeme itself, but the opposite may well be the case for *flower*. Here are the definitions given in *LCED*:

> [10] **embouchure** . . . the position and use of the lips in playing a musical wind instrument
>
> [11] **flower** . . . **1a** a blossom, inflorescence **b** a shoot of a higher plant bearing leaves modified for reproduction to form petals, sepals, ovaries, and anthers . . .

From our discussion so far it will be clear that writing definitions is not a simple, straightforward task with a set of agreed procedures. Besides differences in policy and approach, it appears to be the case that not all words can be defined in the same way. Before we consider this point word-class by word-class, let us review five methods of defining that have been identified. The first is the **analytical** method, which we have discussed in part already. It involves assigning a lexeme to a class of items and then giving details of the individual characteristics of this particular member of the class. The definition of *lynx* at [6] is a good example of the analytical method. Here is another from *LCED*:

> [12] **daffodil** . . . any of various plants with flowers that have a large typically yellow corona elongated into a trumpet shape

First of all *daffodil* is assigned to the class of plants, then we are given details which characterise this plant and distinguish it from other members of the class.

The second method of defining we have also discussed: that of **typifying**, where the definition focuses on what is typical about the referent of the lexeme being defined. We cited the definition of *lyre* at [7] as an example of the typifying definition. Here is a further example from *LCED*:

[13] **hagfish** . . . any of several marine vertebrates that are related to the lampreys, resemble eels, and feed on fishes by boring into their bodies

After being assigned to the class of marine vertebrates, the hagfish is then defined according to three typical characteristics: its relationship to lampreys, its resemblance to eels, and its feeding habits.

The third method of defining is **synthetic**. It may be illustrated with the definition of *green* (noun) in *LCED*:

[14] **green** . . . a colour whose hue resembles that of growing fresh grass or the emerald and lies between blue and yellow in the spectrum

Now, while the definition tells us that *green* belongs to the class of colours, it does not then proceed to analyse *green* by giving us particular details about it. Instead *green* is put into relation with other entities: the colour of typically green objects, and its place in the spectrum of colours. *Green* is in this definition being viewed as part of a whole (the spectrum), rather than being viewed as a whole that has parts (as in the analytical method). It is therefore a synthetic definition rather than an analytical one. We might contrast this definition with the more analytical definition of the colour *yellow* in *CED*:

[15] **yellow** . . . any of a group of colours that vary in saturation but have the same hue. They lie in the approximate wavelength range 585–575 nanometres . . .

Here the wavelength information provides an analysis of the meaning of the colour.

The fourth method of defining is **rule-based** and may be illustrated with the *LCED* definition of *whom*:

[16] **whom** . . . used as an interrogative or relative; used as object of a preceding preposition . . .

In this case, the definition consists of rules expressing how the lexeme being defined is used, i.e. for what purposes and in what contexts.

The fifth method of defining is **synonymy** (see Chapter 5). For example, *daft* is defined in *LCED* as:

[17] **daft** . . . silly, foolish . . .

Two synonyms are given. The assumption must be that if the lexeme is not known to the dictionary user, then the synonym(s) used in the definition will be. Here the principle of defining with words simpler or more widely used than the lexeme being defined is particularly important, but it is not always adhered to. For example, one of the senses of *kind* in *LCED* is defined as 'forbearing, considerate, or compassionate'. But this is an intractable problem for lexicographers: if simple words like *kind* have to be provided with a paraphrase definition then it seems inevitable that more difficult or more complex words will have to be used.

Defining different word-classes

Nouns, especially concrete nouns, may often be defined analytically. The analytical details may refer to size, shape, texture, colour, function, etc. For example, the definition of *lynx* at [6] refers to the size of the tail ('short stubby'), the colour of the coat ('mottled') and the shape of the ears ('tufted'). The definition of *entomology* in *LCED* reads:

[18] **entomology** . . . zoology that deals with insects

Entomology then is a kind of zoology, and its characteristic is that it has the function of dealing with insects. *Entomology* is an abstract noun, which illustrates that these too may be defined analytically. In some dictionaries concrete nouns may be partly defined by means of a pictorial illustration, either a line drawing (e.g. *Webster's Third*) or a photograph (e.g. *Oxford Advanced Learner's Dictionary*): this is a kind of visual analytical or typifying definition. Other nouns, e.g. colours, may be defined synthetically (see the definition for *green* at [14]. And some nouns may be defined by means of synonyms; e.g. the first sense of *genre* is defined in *LCED* as 'a sort, type'. If a noun is derived from a verb or an adjective (e.g. *pollution* from *pollute, possibility* from *possible*), then the definition may include mention of the verb or adjective as a kind of cross-reference for the dictionary user; compare the *LCED* definitions:

[19] **pollution** . . .1 polluting or being polluted 2 material that pollutes

[20] **possibility** . . .1 the condition or fact of being possible 2 sthg possible . . .

Verbs may be defined analytically, especially those referring to observable actions. For example, one sense of *walk* is defined in *LCED* as:

[21] to move along on foot; advance by steps, in such a way that at least 1 foot is always in contact with the ground

Walk is defined as a kind of 'move' or 'advance', with further detail specifying the manner in which it takes place. Many verbs are defined in this way: look up *grab, prattle, refresh.*

The first sense of *grab* is defined in *LCED* as a kind of 'take' or 'seize':

[22] to take or seize hastily or by a sudden motion or grasp

Prattle is defined as a kind of 'chatter':

[23] to chatter in an artless or childish manner

And *refresh* is defined as a kind of 'restore', etc.:

[24] **1** to restore strength or vigour to; revive (e.g. by food or rest) **2** to restore or maintain by renewing supply; replenish **3** to arouse, stimulate (e.g. the memory)

This definition for *refresh* illustrates two further points about the methods of defining verbs. In the first two senses *refresh* is defined both analytically and by means of synonymy (*revive* and *replenish*, respectively), and the third sense is defined wholly by synonyms. Synonymy is frequently used as a method for defining verbs, and some verbs are defined only by means of synonyms, e.g. *deride* in *LCED*:

[25] to mock, scorn

The other feature of verb definitions illustrated by *refresh* in [24] is the use of typical collocations; e.g. in the first sense, some typical means of refreshing (= reviving) are suggested, and in the third sense a typical object is suggested.

Adjectives may be defined by a number of methods. We find an analytical definition, supported by a synonym, in the definition of *parsimonious* in *LCED*:

[26] frugal to the point of stinginess; niggardly

where it is defined as a kind of 'frugal'. Where adjectives relate to or are derived from nouns or verbs, their definitions often refer to these; e.g. *parochial* in its first sense is defined in *LCED* as:

[27] of a (church) parish

Similarly, *retentive* is referred to *retain*:

[28] able or tending to retain; esp retaining knowledge easily

Many adjective definitions begin with the present or past participle form of a verb; e.g. the first two senses of *patient* are defined in this way in *LCED*:

[29] **1** bearing pains or trials calmly or without complaint **2** manifesting forbearance under provocation or strain

The verb + object ('bearing pains or trials') is what *patient* is a kind of, and the further detail is the manner in which it is done; so that this is in fact a kind of analytical definition. As with other word-classes, adjectives are sometimes defined merely by a synonym, e.g. the first sense of *paternal* in *LCED*:

[30] fatherly

Adverbs (of manner) which are derived from adjectives by addition of the suffix *-ly* do not usually receive separate definition; they are listed as derivations (run–ons) under the appropriate adjective. However, look up the adverbs *frankly* and *hopefully* in your dictionary.

In both *LCED* and *CED* these adverbs are entered as separate headwords. *Hopefully* has two senses listed in both dictionaries:

[31] **1** in a hopeful manner **2** it is hoped

The first sense is the manner adverb sense. The second sense, which is marked 'informal' in *CED* and 'disapproved of by some speakers' in *LCED*, is the reason for the separate entry of this adverb. In *CED*, *frankly* also has two senses:

[32] **1** (sentence modifier) in truth; to be honest . . .
2 in a frank manner.

The second sense is the manner adverb, and the first, sentence modifier, sense is the reason for the separate entry. The *LCED* entry has only one sense listed; it omits the manner adverb sense. Other adverbs that have multiple senses or are not derived from adjectives receive separate entries, e.g. *away, fast, soon*. They are frequently defined by means of synonyms, or by means of prepositional phrases that include a noun referring to the kind of circumstance denoted (e.g. time, manner, place); e.g. *soon* has the definition in *LCED*:

[33] without undue time lapse, in a prompt manner

Grammatical words (determiners, pronouns, prepositions, conjunctions) are sometimes defined by means of synonyms, but frequently their definitions are framed in terms of rules of use. The first sense of *the* in *LCED* reads:

[34] used before nouns when the referent has been previously specified by context or circumstance

And the first sense of the preposition *in* reads:

[35] used to indicate location within or inside sthg three-dimensional

In the *LCED* account of the first sense of the preposition *after* the definition is partly analytic and partly rule-based:

[36] behind in place or order . . . used in yielding precedence . . . or in asking for the next turn . . .

After is defined first of all as a kind of 'behind', and then we are given the rules for using it in this sense.

Beyond definition

We suggested at the beginning of this chapter that the function of dictionary definitions was to express the meaning of each (sense of each) lexeme distinctively. This does not mean that dictionary definitions have to correspond with scientific definitions or that they must be comprehensively analytical. The important function of a dictionary definition is that it typifies the meaning of a (sense of a) lexeme. In this sense a dictionary is different from an encyclopaedia, though it must be said that the dividing line between them is by no means clear. Dictionaries become like encyclopaedias in two respects.

First of all, dictionaries – some more than others – contain entries that are more associated with the encyclopaedia than with the dictionary, especially proper names of people, places, institutions, etc. A glance through your dictionary to note the number of headwords beginning with a capital letter will reveal the extent to which it includes proper names. Most dictionaries, including *LCED*, include proper names of places (usually countries) or peoples (nations, cultural groups) or institutions that have some degree of general currency. A couple of pages in *LCED* include for example:

[37] Manx   Manzanilla   Maoism   Maori   Maori-
     tanga   Maratha   Marathi   March   Mardi-Gras

Strictly speaking, some of these words are not proper nouns, but either adjectives (e.g. *Manx*) or common nouns (e.g. *Maoism*) derived from proper nouns. If you do not know the meaning of any of these, you are advised to look them up in your dictionary. *CED* has a considerably more extensive range of proper names, including not only countries but towns and cities also, as well as famous people past and present. Between *Manx* and *Mardi Gras*, for example, there are twenty-seven proper names in addition to those listed at [37] for *LCED*. But this is a feature of *CED* and not found in most British dictionaries until recently (though more commonly in American dictionaries).

The other way in which dictionaries approach encyclopaedias is when their definitions become encyclopaedic. This is particularly the case in respect of phenomena (e.g. plants, animals) that have been subject to scientific classification and description. But dictionaries vary in the extent to which they become encyclopaedic in the definition of these lexemes. *CED*, for example, goes further in this direction than *LCED*. Compare the entries for *badger*:

[38] *LCED*:   '(the pelt or fur of) any of several sturdy
                burrowing nocturnal mammals widely
                distributed in the northern hemisphere'
[39] *CED*:   'any of various sturdy omnivorous
                musteline mammals of the subfamily

> *Melinae,* such as *Meles meles* (Eurasian badger), occurring in Europe, Asia, and North America: order *Carnivora* (carnivores). They are typically large burrowing animals, with strong claws and a thick coat striped black and white on the head.'

In the *CED* definition, not only are technical terms used, such as *musteline,* and the scientific classification given, but an attempt is made to give a detailed picture of the badger. Whether such detail is included in dictionary definitions, depends on the group of users that the dictionary is aimed at. We would not expect to find it in dictionaries for foreign learners, for example.

Exercises

1. How many senses of *see* would you identify on the basis of the following examples?
   (a) I can't scc Lydia anywhere.
   (b) The security guard asked to see our passes.
   (c) Can you see what I mean?
   (d) We're going to see a film tonight.
   (e) I could see that you were having an argument with him.
   (f) I'd like to see the manager, please. I have a complaint.
2. Are the senses of lexemes in your dictionary ordered historically or by supposed commonness of usage? Check in the front-matter of the dictionary and see which sense comes first in the entry for *scapegoat.*
3. Look up the following words in your dictionary and say which method of defining has been used (analytical, synthetic, synonym, rule-based).
   here   milk   humdrum   hundred   hypermarket
4. Look up the following words in your dictionary. Has any of them been given an encyclopaedic definition?
   herring   hexameter   hilum   histamine   hollyhock
5. Write a definition for the lexeme *holiday.* How many senses will you identify? What method of defining will you use? Check your attempt against two or more dictionaries.

CHAPTER 10

More than Meaning

At the beginning of the previous chapter we noted that it is information about the meaning of words that we most expect to find in a dictionary. And yet there is at the same time an expectation that the dictionary will tell us how to 'use' words correctly. Clearly that involves knowing the meaning of a word so that it can be used appropriately, in the context of other words; but it must also involve knowing what the grammatical possibilities of a word are, as well as the situational contexts for which the word is appropriate. If we also take account of the recording function of dictionaries, then it is not surprising that they contain much more than just definitions. In this function we need to see the dictionary in relation to other descriptive works about language, especially grammars. At the level of word, grammars describe how classes (or subclasses) of words operate in the grammatical structure of language, whereas dictionaries treat words as individual units and describe how they operate idiosyncratically in the language.

Dictionaries contain therefore a considerable store of information about the grammar and usage of lexemes, not to mention information on spelling (inevitably), pronunciation and etymology (compare Chapter 3). Examine the following entry for *rumble* (verb) from the *Longman Concise English Dictionary* (*LCED*). What information does it give besides the definitions?

> [1] ¹**rumble** /'rumbl̩/ *vb* **rumbling** /'rumbling, 'rumbl.ing/ *vi* **1** to make a low heavy rolling sound **2** *NAm* to engage in a street fight – infml ~ *vt* **1** to utter or emit with a low rolling sound **2** to reveal or discover the true character of – infml [**ME** *rumblen*; akin to **MHG** *rummeln*, to rumble] – **rumbler** *n*

Pronunciation is indicated in the slashed brackets, using *LCED*'s own notation system based on the Roman alphabet, including the stress pattern. *Rumble* has main (primary) stress on the first syllable, with the second syllable unstressed. We are then informed that rumble belongs to the word-class (part-of-speech) of verbs, and that the present participle is formed (irregularly) by dropping the final *e* in spelling, which is then pronounced either as a two-syllable word /'rumbling/ or as a three-syllable word /'rumbl.ing/. Next we notice that *rumble* belongs both to the subclass of intransitive verbs (vi) and to that of transitive verbs (vt), each of which has two senses. The second sense of the intransitive use is restricted geographically to North American English. And this sense, together with the second sense of the transitive use are restricted stylistically to 'informal' usage. The square brackets contain etymological information, which traces the origin of *rumble* to Middle English and notes a cognate word in Middle High German. Finally, the entry notes a regular noun derivation from *rumble* by means of the *-er* suffix.

This entry for *rumble* contains a wide diversity of information additional to the definitions. In this chapter we are going to examine especially the grammar and usage information contained in dictionary entries. This will expand on some of the points touched on towards the end of Chapter 3.

Grammatical words

In Chapter 1 we made a distinction between **'lexical'** words and **'grammatical'** words, where we noted that members of the grammatical word-classes, such as determiner or pronoun, have a predominantly language-internal function, making their contribution to the grammar of sentences rather than to their referential meaning. We might therefore expect such words to be described fully in a grammar, rather than in a dictionary. And that is indeed the case; for example, the many and various uses of the definite article *the* are described fully in a grammar such as Quirk *et al.*, *A Comprehensive Grammar of the English Language* (1985). But you will also find *the* treated in some detail in desk-size dictionaries like the *Longman Concise English Dictionary*

(*LCED*) or the *Collins English Dictionary* (*CED*), though it is doubtful whether any native speaker, apart from linguists or dictionary students, would ever consult a dictionary in order to find out about *the*.

There is an overlap between grammar and dictionary at this point. The dictionary in its recording function needs to treat all words, including those that are fully described in the grammar. We noted in Chapter 9 (p. 139) that such words often have rule-based definitions in the dictionary: these rules are then of a grammatical nature, indicating how a grammatical word functions in the structure of the language. But grammatical words are not all or exclusively defined in this way, and this confirms our contention in Chapter 1 (p. 17) that there is a gradation from 'fully lexical' to 'fully grammatical'. Look up the following 'grammatical' words in your dictionary and notice whether they are defined solely by rules or in some other way as well or instead:

[2] me   their   yours   which   with   if   because
    might (past of *may*)

Me is a personal pronoun, the object (or objective) form of *I*; in *LCED* it is defined solely by means of a rule; in *CED* an attempt is made at a paraphrase definition, but with an extensive 'usage' note. *Their* is a possessive determiner: it is defined in *LCED* by both paraphrase and rule, but in *CED* by paraphrase only. *Yours* is a possessive pronoun; it is defined in both *LCED* and *CED* partly by paraphrase and partly by rule. *Which* may function as a relative pronoun or as a relative/interrogative adjective/determiner: *LCED* defines by a combination of paraphrase and rule, *CED* mainly by rule. *With* is a preposition; it is defined in both *LCED* and *CED* by a combination of paraphrase and rule. *If* is a conjunction, and it is likewise defined partly by paraphrase and partly by rule. *Because* is also a conjunction; it is defined solely by paraphrase. *Might* belongs to the closed (grammatical) subclass of modal verbs, and it is defined in both *LCED* and *CED* solely by rule. The method of definition in each case would appear to indicate whether the item is used only grammatically (e.g. *me*, *might*), both grammatically and lexically (e.g. *their*, *with*, *if*);

or only lexically (e.g. *because*). It is therefore not surprising that, in view of their often indeterminate status, so-called 'grammatical' words should be treated in the dictionary, in fulfilment of its recording function.

Word-class

We noted that in [1] the entry for *rumble* contained an indication of its word-class or part-of-speech, namely verb (vb). This has been a traditional piece of information from some of the earliest dictionaries. It may well be a practice carried over from Latin dictionaries: this is crucial information for Latin words, since knowing the part-of-speech is essential for working out the correct inflections. This could also be a justification for including word-class labels in English dictionaries: indicating that a word is a noun implies that it is likely to inflect for plural number (unless it is 'uncountable') and in some instances for possessive (genitive) case (e.g. *fields, field's*).

However, inflectional information is not the only kind that is implied by a word-class label. It also implies information about the syntactic operation of a word. A noun, for example, may occur in certain positions, fulfil certain functions, in the syntactic structure of a language. Being a noun determines the kind of syntactic relations that a word may enter into, e.g. modifier – head relation with an adjective ('green – fields'), subject – predicator relation with a verb ('the fields – rejoice'). Broad word-class divisions are, though, a rather rough guide to syntactic operation, and it is not clear that lexicographers intend them to be that; perhaps they are only following a tradition in including them. Another tradition may be the practice of subclassifying verbs in particular into **transitive** and **intransitive**; but this does make the syntactic information a little less crude. 'Vi' tells us that the verb does not occur with an object (e.g.'This machine is not working'), 'vt' that it does (e.g. 'The farmer is ploughing **his field**'). Some dictionaries, e.g. those in the Oxford family, add a third subclass label, 'absol.' (= absolute), referring to the use of a transitive verb but without an object stated (e.g. 'They are reading'). Look at the front-matter of your dictionary and at the entry for *read* to discover what word-class labels it assigns to verbs.

LCED has three labels: 'vb', 'vi' and 'vt', all illustrated in the entry for *rumble* at [1]. 'Vb' appears to be used only if some information intervenes between the pronunciation and the first sense. *CED* has three labels: 'vb.', 'tr.' and 'intr.'. All members of the word-class are marked 'vb.'; if a sense is exclusively transitive it is marked 'tr.', if exclusively intransitive 'intr.'; and if it may be used either transitively or intransitively it is not marked at all. We may question whether this subclassification is of any use to users of monolingual English dictionaries, either on the argument that native speakers do not learn the syntactic use of words from the dictionary but from experiencing them in speech and writing, or because a really useful subclassification (for foreign users if not for native speakers) would be more finely differentiated, as for example in the *Longman Dictionary of Contemporary English* (see Chapter 12, p. 179).

Look up the following nouns in your dictionary. Is any further information provided for any of the senses beyond the word-class label 'n.'?

[3] garden   meat   wine

One subclassification that is of some syntactic significance is that into **countable** (i.e. can be made plural and be counted, e.g. 'six boxes') and **mass** (**uncountable**) nouns (i.e. cannot be counted, e.g. 'some money'), but dictionaries do not usually indicate·it. Of the lexemes in [3], *garden* would be marked 'countable' and the other two 'mass', except that *wine* may be countable with the special meaning 'kinds of' (e.g. 'six wines from Germany'). The distinction is significant because it affects the determiners that may be used with a noun. Countable nouns (since they may have plural forms) may combine with quantifiers like *many* and *several*, numerals, and the indefinite article *a* (e.g. 'three gardens', 'several reports'). Mass nouns, on the other hand, have only a singular form and combine with quantifiers like *some* and *a lot of* (e.g. 'some furniture', but not *'a furniture' or *'some furnitures'). Another subclass of nouns that has syntactic significance contains those which may function like adjectives as modifiers of another noun; e.g. *garden*, as in 'garden party', 'garden furniture'. *CED* marks (senses of) nouns which regularly occur in this function with the label 'modifier'.

Now look up the following adjectives in your dictionary, to establish whether any subclassification is marked for any of the senses:

[4] asleep   ill   mere

The significant subclassification here is that into **attributive** adjectives (occurring before a noun, e.g. 'a big farm') and **predicative** adjectives (occurring after a verb like *be*, e.g. 'the farm is big'). Most adjectives belong to both subclasses, like *big*; but some are found in only one. *Asleep*, for example, functions predicatively only; *mere* attributively only; and *ill*, in the sense of 'sick', is usually used only predicatively, though some speakers use it attributively as well. *CED* marks adjectives like *mere* with the label 'pre-nominal', those like *asleep* with the label 'postpositive', and it marks *ill* as 'usually postpositive'.

Finally look up the following 'adverbs' in your dictionary. What word-class label are they given, and is there any subclassification?

[5] personally   possibly   therefore   very

The class of adverbs contains a number of quite diverse subclasses, but most dictionaries mark all the words in [5] merely as 'adv.'. *Personally* belongs, in one of its senses, to a subclass of 'viewpoint' adverbs, in sentences like 'Personally, I think it's a good idea': *CED* marks this sense with the label 'sentence modifier'. The same label is used for one of the senses of *possibly*; another has the label 'sentence substitute', to indicate the use of *possibly* to stand for a whole sentence (e.g. 'Will you call in tomorrow?' – 'Possibly'). *Therefore* belongs to a subclass of 'conjunctive' adverbs; it serves to connect a sentence to a preceding one (e.g. 'We have no money. We therefore cannot afford a holiday.'): it is marked in *CED* with the label 'sentence connector'. *Very* belongs to a subclass of 'intensifying' adverbs; it functions as a modifier for adjectives and other adverbs (e.g. 'very big', 'very quickly'): *CED* marks it with the label 'intensifier'.

From our discussion it will be clear that *CED* makes more subclassifications of word-classes than do most monolingual dictionaries, though this is a feature that the foreign learners' dictionaries pay considerable attention to (see Chapter 12, p. 179). One further point needs to be made about word-class labels. It will have become clear from the examples investigated that word-class and subclass membership is used by dictionaries as one of the criteria on which sense divisions of lexemes are made. If you look back to the entry for *rumble* at [1], you will notice that it is in fact *1rumble*, because there is a *2rumble*, which is the noun. *CED*, as we have noted before (Chapter 3, p. 41), does not have separate headwords in such cases, but uses the word-class division as a basis for the division of (groups of) senses under the one headword. Within the *LCED* entry for *1rumble*, though, you will notice that a division into groups of senses has been made on the basis of 'vi' and 'vt'. This, then, is a further function of the word-class label in dictionary entries.

Inflections

Most dictionaries give information on the irregular inflections of words, i.e. those that do not conform to the majority pattern, such as the addition of the -(*e*)*s* suffix for noun plurals or the -(*e*)*d* suffix for verb past tense and past participle. Look up the following words in your dictionary and note the inflectional information given:

[6] bad louse run show.

Bad is one of a small number of adjectives which inflect irregularly for comparative and superlative degree, and dictionaries will usually list the forms *worse* and *worst*. *Louse* is a noun which forms the plural not by the addition of the suffix but by a change of vowel internally, with consequences for the spelling: *lice*. Your dictionary should also note that there is one sense of *louse* ('a contemptible person') for which the plural is formed according to the regular pattern. One further point about noun inflections is that some nouns exist only in the plural form, e.g. *scissors, trousers*; and this too is usually indicated in the dictionary entry, e.g. by the abbreviation 'n.pl.'.

Most of the irregular inflections of nouns (of Anglo-Saxon origin) occur with commonly used words (*foot, mouse, tooth*, etc.), and native speakers would not normally need to consult a dictionary for this information. However, there is a set of nouns, about the plurals of which many native speakers are confused or uncertain. These are the nouns of Latin or Greek origin, which have retained their original plural forms. In some cases the plurals have been regularised on the pattern of English and exist side by side with the Latin or Greek plurals. Look up the following nouns in your dictionary and note the plural forms given:

[7] criterion   index   addendum   datum   data

Criterion derives from Greek, and its plural is given in *LCED* as '*criteria*, also *criterions*'. *CED* adds a usage note to point out that the use of the (plural) form *criteria* as a singular 'is not acceptable . . . in careful written and spoken English'. *Index* derives from Latin, and its plural is given as '*indexes, indices*', with the regularised English plural first (compare *criterion*). *Addendum* also derives from Latin; its plural is given as *addenda* only; but the *LCED* entry notes that it is 'often pl with sing. meaning but sing. in constr[uction]', e.g. in a sentence such as 'The addenda to this book is the most interesting part'. In the case of *datum* and *data* the separation of plural from singular has gone so far that *data* merits a separate entry in the dictionary. In *LCED* the plural of *datum* is given as *data* for one of its senses ('sthg given or admitted, esp as a basis for reasoning or drawing conclusions'), and as *datums* for the other ('sthg used as a basis for measuring or calculating'). *Data* is marked as 'pl but sing. or pl in constr'; i.e. it is used both in a sentence like 'The data **has** been examined' and in one such as 'The data **have** been examined'.

Let us return now to the verbs in [6]: *run* and *show*. Most irregular inflections of verbs in English are associated with the past tense and past participle forms. *Run* forms past tense by means of an internal vowel change (*ran*), while the past participle (irregularly) has the same form as the base/present tense. In the case of *show* the only irregular form is the past participle, which has retained the Middle English −*n* suffix in preference to the modern English −(*e*)*d* suffix, though, as the dictionaries indicate, the latter is also now used. *LCED* gives only the form of the verb which is inflected irregularly, while *CED* gives all the forms if any one of them has an irregular inflection: so *LCED* gives only the past participle form of *show*, while *CED* gives the third person singular present tense (*shows*), present participle (*showing*), past tense (*showed*), as well as past participle (*shown/showed*).

Dialect and time

The information about lexemes that we have been considering so far in this chapter has been grammatical. We now turn to information about usage, which we shall examine under a number of headings. We are concerned with ways in which language varies, and the first parameter of variation to be considered is geographical. Look up the following words in your dictionary. Are they marked as geographically restricted?

> [8] day-return decedent decoke destruct dinkum
> divot donga doolan dorp dreich drongo
> dustman

All these words have geographical labels in *LCED*: *day-return*, *decoke* and *dustman* are marked 'Br' (i.e. British); *decedent, destruct* are marked 'NAm' (i.e. North American); *dinkum, drongo* are marked 'Austr' (i.e. Australian); *divot, dreich* are marked 'Scot' (i.e. Scottish); *donga, dorp* are marked 'SAfr' (i.e. South African); and *doolan* is marked 'NZ' (i.e. New Zealand). These words are then marked as restricted to particular national varieties of English.

Now look up the following words in your dictionary and note any geographical label. You may not find some of these words in your dictionary; they are all in *CED*, however.

[9] butty canny canty carzey charm (noun) chaw
    chelp chine chollers chumping clarts cleck

These words are all marked in *CED* as restricted to a regional dialect of British English. Dictionaries vary though in the number of dialect words that they include and in the detail of the labels used to mark them. For example, *LCED* includes only four of the words in [9], and for *butty* (= 'sandwich') notes merely that it is 'dial Br' (i.e. British dialect), whereas *CED* labels *butty* as 'chiefly northern English dialect'. *Canny* occurs in both dictionaries, labelled 'Scot. and northeast English dialect', but the dictionaries differ in the definitions: *LCED* gives the dialect senses as '1 careful, steady' and '2 agreeable, comely'; while *CED* gives '1. good, nice: used as a general term of approval' and '2. quite, rather'. *Canty* (= 'lively') is labelled 'Northern Brit.' in *CED*, *carzey* is 'Cockney slang' for 'toilet', *charm* is 'Southwest Brit.' for 'a loud noise such as people chattering or birds singing'. *Chaw* occurs in both dictionaries, marked 'dialect', and is defined as 'chew (tobacco)'. *Chelp* occurs in *CED* with the dialect label 'Northern and Midland English' and is defined as '(esp. of women or children) to chatter or speak out of turn'. *Chine* (= 'deep fissure or steep ravine') is marked in *CED* as 'Southern Brit. dialect', whereas this information is included in the definition in *LCED*: 'esp in Dorset or the Isle of Wight'. *Chollers* (= 'jowls, cheeks') occurs in *CED* with the label 'Northeast English dialect', *chumping* (= 'collecting wood for bonfires on Guy Fawkes Day') is 'Yorkshire dialect', *clarts* (= 'lumps of mud') is 'Northern English dialect', and *cleck* (= 'gossip, tell on') is 'South Wales dialect'.

The lexemes in [8] and [9] are marked as belonging to particular geographical varieties of English. Other lexemes are marked as belonging to particular temporal varieties. Two labels are commonly employed for this purpose: 'obsolete' and 'archaic'. 'Obsolete' means that a lexeme, or sense of a lexeme, is no longer in use; in *LCED* it means specifically that there is no evidence of use since 1755 (the date of Johnson's dictionary). 'Archaic' means that a lexeme, or sense of a lexeme, is no longer commonly used, but may still be found in literary and other writing to impart an historical flavour. Look up the following words in your dictionary and note any temporal label:

> [10] fain fluxions footpad fox (= 'befuddle with alcoholic drink') froward gentle (= 'ennoble') gesture (= 'posture') glaive glory (= 'brag') groom (= 'manservant')

Again you may not have found all these lexemes or senses of lexemes in your dictionary. The following are marked as 'obsolete' in *CED*, though they do not occur in *LCED* (with these senses): *fox, gentle, gesture, glory*. From this list it will be clear that general dictionaries do not usually include lexemes which are obsolete, but only a current lexeme's senses which are obsolete. Archaic lexemes are included, however, since they may still be encountered by dictionary users. The remaining words, or senses of words, in [10] are marked 'archaic' in either or both *CED* and *LCED*.

Formality and status

We turn now to information given about words in respect of the social context in which they are usually used. We are concerned on the one hand with formality of context, where words are labelled especially 'informal' or 'colloquial' if they are restricted to that kind of context. A smaller number of words is labelled 'formal' if they are restricted to such contexts, e.g. the legal words *hereunder, heretofore*, etc. The great majority of words may of course be used in any context for which their referent is appropriate, e.g. we would expect technical terms to be naturally excluded from informal contexts. On the other hand we are concerned with the status of words in terms of their disapproval by the speech community at large and therefore the restriction on the contexts in which they are used. Labels marking this kind of restriction include 'slang', 'vulgar', 'taboo'. Clearly the two kinds of social restriction are related. Now look up the following words in your dictionary and note any social labels of formality or status:

> [11] arse bum (noun) cussed fart git hot air

piss   pissed (= 'drunk')   scram   shit   tit
(= 'breast')   woozy

If you have been able to consult more than one dictionary, you will have found that they do not always agree on how to mark the words in [11]. For example, *LCED* marks *arse* as 'vulg', while *CED* uses 'taboo'; though these two terms seem to correspond. *Bum* is marked as 'slang' in both, as is *git*. *Cussed*, along with *hot air, scram* and *woozy* are marked 'informal' in both dictionaries. *Fart* has the 'vulg' label in *LCED* and the 'taboo' label in *CED*. *Piss* is likewise marked 'vulg' in *LCED*, but 'taboo slang' in *CED*; *pissed* has the same label in *CED*, but in *LCED* it is marked merely 'slang'. *Shit*, in its primary sense, is not marked at all in *CED*, but as 'vulg' in *LCED*. For *tit*, however, *LCED* has the label 'infml', against *CED*'s 'slang'. Clearly the differing opinions of dictionary editors reflect the range of attitude in the speech community at large.

Also under this heading we need to mention items like the famous *ain't*, which we discussed in connection with *Webster's Third New International Dictionary* in Chapter 8, p. 123. In *LCED* and *CED*, *ain't* is labelled 'chiefly nonstandard' and 'not standard' respectively. It offends, therefore, not so much against the rules of linguistic decency, like the 'vulg' or 'taboo' words, but rather against the linguistic social graces. Similarly marked is *what* in its relative pronoun function ('That's the man what I saw yesterday'), obtaining the label 'substandard' in *LCED* and 'not standard' in *CED*. One further example we could cite in this regard is *never*: *LCED* labels the usage 'No, I never' (in answer to a question, 'Did you . . .?') as 'nonstandard'; and *CED* in a usage note points out that 'in good usage *never* is not used with simple past tense to mean *not*' (e.g. 'I never heard of it').

Domain

There is one further set of labels that is frequently used in dictionaries. They mark the domain or field of discourse to which a word may be restricted, e.g. 'Botany', 'Nautical',

'Music'. The labels inform the dictionary user that such lexemes (or senses) do not belong to the common core of the vocabulary, but to a specialist or technical lexical field. Look up the following lexemes in your dictionary and note any labels marking domain. Some dictionaries, however, include such information as part of the definition rather than by means of a special label.

[12] allegro   chantry   dramatic irony   foreclose
hypotaxis   jarl   macula   northing   placket
resupinate   solvate   visual display unit

The labels, all different, which we shall give for the domains of these lexemes all come from *CED*; *LCED* appears not to mark domains, though the domain is usually indicated in the definition. This can be illustrated with *allegro*, which is defined in *LCED* as:

[13] (a musical composition or movement to be played) in a brisk lively manner

whereas *CED* labels the lexeme 'Music' and defines it as:

[14] **1.** quickly; in a brisk lively manner. . . **2.** a piece or passage to be performed in this manner

The domain is not always indicated in *LCED*, however. In the case of *resupinate*, *CED* marks as 'Botany' and defines as:

[15] (of plant parts) reversed or inverted in position, so as to appear to be upside down

LCED defines *resupinate* merely as:

[16] (appearing to be) upside down

Returning to the list in [12], *chantry* is labelled 'Christianity' in *CED*, *dramatic irony* 'Theatre', *foreclose* 'Law', *hypotaxis* 'Grammar', *jarl* 'Medieval history', *macula* 'Anatomy', *northing* 'Navigation', 'Astronomy' and 'Cartography' for different senses, *placket* 'Dressmaking', *solvate* 'Chem[istry]', *visual display unit* 'Computer technol[ogy]'. This by no means exhausts the range of labels for domains in the *CED*, though nowhere in the front-matter is a list of domain labels given, such as you can find, for instance, in the *Oxford*

Advanced Learner's Dictionary of Current English (see Chapter 12, p. 177).

Quotations and examples

We shall consider one further item additional to the definition, which appears in some entries in some dictionaries: the quotation or example. In an historical dictionary like the *Oxford English Dictionary* each word is supplied with quotations which are acknowledged (the author or source is given) and dated; their function is to provide an illustrated history of the meaning of the words, and to authenticate the definition. In a general-purpose desk-dictionary the quotations or examples do not have this function. Mostly such dictionaries contain examples rather than quotations; that is, phrases or sentences provided by the lexicographer, rather than cited from a written source. They are not usually provided for every lexeme or for every sense of a lexeme; *LCED*, for example, tends to provide them for lexemes with multiple senses. And for *LCED* the purpose of the examples is to illustrate 'a typical use of the word in context'; they serve therefore to reinforce the definition and to show typical usage. They are often used to illustrate a lexeme or sense of a lexeme that occurs in only a limited range of contexts; e.g. in both *LCED* and *CED extraditable* is illustrated by the phrase 'an extraditable offence'. For some dictionaries, the division into senses will be made on the basis of the examples collected or thought of. This appears to have been the case with *runaway* (adjective) in *LCED*:

> [17] **1** fugitive **2** accomplished as a result of running away <*a~ marriage*> **3** won by a long lead; decisive <*a~ victory*> **4** out of control <*~ inflation*>

Let us take as a detailed instance of the use of quotations and examples the entry for *quality* in *LCED*. You may like to compare this with the entry for this lexeme in your dictionary. The meaning of *quality* is divided into six senses in *LCED*, the first two being further subdivided into two each. Sense **2a** ('degree of excellence; grade') is illustrated by an example, 'a decline in the ~ of applicants'. Sense **2b** ('superiority in kind') is illustrated by a quotation from Compton Mackenzie, 'proclaimed the ~ of his wife'. Sense

3 ('high social position') is illustrated by an example of typical context, 'a man of ~'; as is sense 4 ('a distinguishing attribute; a characteristic'), 'listed all her good ~'. Finally, sense 6, which is marked 'archaic', ('a capacity, role') is illustrated by a quotation from Joseph Conrad, 'in the ~ of reader and companion'. It is for words like *quality*, which have quite complex meanings, and where the definitions may not be sufficiently explanatory by themselves, that examples or quotations are particularly appropriate.

Exercises

1. Look up the entries for *us* and *them* in your dictionary. How are these words defined? What additional information are you given?
2. What grammatical information is supplied in your dictionary for the following lexemes?
   cloth   have   little   memorandum   up
3. Examine the entries from *jail* to *jar* in your dictionary and note any usage labels employed.
4. Examine carefully the entry for *speak* in your dictionary. Do the quotations/examples help you to understand the definitions of the senses for which they are provided? Are the definitions without examples harder to understand? You may like to invent examples for the latter.
5. Without consulting a dictionary, make up your own desk-dictionary entry for (say) *tranquil, frontier* and *kid* (verb and noun). If appropriate you may like to work together with other students and compare your attempts. Then compare what you have written with the corresponding entry in an actual dictionary. How is it different? In what respects is your entry better or worse than that in the dictionary?

CHAPTER 11

Different Dictionaries

The term 'dictionary' appears in the titles of a wide range of reference works, many of which are not strictly lexicographical. For example a 'Dictionary of Quotations' is usually a listing of famous or frequently-used quotations in alphabetical order based on a key word or words in each quotation. The term 'dictionary' is presumably used because the listing is alphabetical and there is some connection with words, but there is no lexicographical description as such. In looking at different dictionaries in this chapter we shall restrict ourselves to those that are lexicographical. A dictionary will therefore be taken to be an alphabetical listing of words with descriptive information about them, intended to be used for reference purposes. This definition still encompasses a wide range of publications, as you will see if you look at the appropriate shelves of your local bookshop.

A commercial product

When we critically consider a dictionary we have to remember that it is a commercial product. A publisher has considered it to be financially worth investing in the editorial manpower and time needed to produce a dictionary in the expectation that it will more than pay for itself in subsequent years once it is published. In most cases this means maintaining a dictionary department which will not only work on future editions of dictionaries, but will also continually update the files of material on which dictionary entries are based.

As commercial products, dictionaries fulfil a market

need. To some extent that need is given. We suggested at the beginning of Chapter 3 (p. 35) that, as with the Bible, there is the assumption that most English households will have a dictionary, whether – again like the Bible – it is referred to or not. The education system reinforces the need for a dictionary; and as we have noted before, 'the dictionary' is regarded as the repository of the language by many people. Where this is the case, and we are talking mainly about general-purpose dictionaries (see below, p. 159), publishers in the marketing of their dictionary try to emphasise some feature which makes their product stand out by comparison with its rivals in the marketplace. Look at your dictionary and identify what the publishers emphasise as the striking feature or features.

The *Longman Concise English Dictionary* claims to be 'unrivalled for the size in its coverage of today's English' and to be 'the first dictionary of its size to be based on evidence gathered from many hundreds of books, periodicals, newspapers and journals to be sure that it reflects the current state of the English language' (from the dust jacket). The *Collins English Dictionary* claims to be a 'major new dictionary', containing 'more vocabulary references and more text than any comparable one-volume dictionary', setting 'new standards both in the extent of its coverage and the clarity of its presentation' (from the dust jacket). You will notice the implicit ('unrivalled') or explicit ('more . . . than any comparable one-volume dictionary') comparisons that are being made. The market for general-purpose dictionaries exists: there is competition by each dictionary for its market share.

In the case of the more specialist dictionaries (see below) a market needs to be identified or even created, but the conditions are generally more akin to ordinary textbook publishing. Specialist dictionaries are often the work of one person working on their own or are derived from the publisher's general-purpose dictionaries. Occasionally, and supremely in the case of the *Oxford English Dictionary* (*OED*), a publisher will have to invest in a dictionary enterprise with no immediate prospect of great financial return. In the case of the *OED*, where nobody anticipated the scale of the undertaking, this led to constant friction between editor and publisher, as Katharine Murray so ably recounts in the biography of her grandfather, James A. H. Murray *Caught in the Web of Words*, (1977). But the *OED* is primarily a work of historical scholarship, not a dictionary in the commercial sense. There is perhaps a tension between the ideals of lexicographers and the commercial demands of publishers, which surfaced most acutely in the case of the *OED*. This is a point we will return to at the end of the chapter.

We have implied in our discussion so far that there are two broad categories of dictionary: general-purpose dictionaries and specialist dictionaries. General-purpose dictionaries are intended to contain all the lexicographic information that users might want to look up. The existence of specialist dictionaries implies that either there is information which general-purpose dictionaries do not, or do not adequately, deal with, or there are groups of users who are not served adequately by general-purpose dictionaries. Perhaps general-purpose dictionaries are attempting the impossible. Are they locked into a lexicographical tradition (of comprehensively recording the language) that no longer really serves the linguistic complexity of the modern world? We shall, hopefully, be able to shed some light on these questions as we consider the different dictionaries that are on offer to today's user.

General-purpose dictionaries

General-purpose dictionaries are what most of us buy and what we conceive of as 'the' dictionary. They are the dictionaries that are the main object of consideration in this book. They contain an alphabetical listing of the vocabulary and aim to give a comprehensive coverage of the vocabulary within the limits of their size. They are compiled within a lexicographical tradition that defines lexical information about words as consisting of at least: pronunciation, irregular inflections, part-of-speech (word-class), definitions, etymology, stylistic and dialectal restrictions, and possibly field of use. We have discussed these constituent parts of dictionary entries in some detail in Chapters 3 and 10, and we have questioned the appropriateness or usefulness of some of this information for many dictionary users. But we have to recognise that dictionaries are not only commercial products, they are also the product of nearly four centuries of the developing lexicographical tradition (as we have traced it in Chapter 8); so that for a dictionary to be a dictionary it must be seen to be located within that tradition. We shall return to this point at the end of the chapter.

Meanwhile, let us note that general-purpose dictionaries are available in different sizes. Most publishers produce a range of dictionaries, aimed at slightly different corners of the market and priced accordingly. Discounting the very large, essentially library dictionaries like *Webster's Third*, dictionaries for general use come in three or four sizes. Largest of the range are the 'desk-dictionaries' such as the *Collins English Dictionary (CED)*, the *Longman Dictionary of the English Language* or *Chambers Twentieth Century Dictionary*. Next in size come the 'concise' dictionaries such as the *Concise Oxford Dictionary of Current English*, the *New Collins Concise English Dictionary* or the *Longman Concise English Dictionary*. At the smallest end of the range come the 'pocket' or 'compact' dictionaries, such as the *Pocket Oxford Dictionary*, the *Collins Pocket English Dictionary* or the *Longman Compact English Dictionary*. Some publishers (e.g. Collins) have both a 'pocket' and a still smaller 'compact' edition of their dictionary; indeed, Collins have a yet smaller 'gem' dictionary, making a range (from *CED* to *Gem*) of five dictionaries in all.

To ascertain how these dictionaries of different sizes differ in their content we will examine the 'desk', 'concise' and 'pocket' editions in the Collins range. The desk-size *CED* has a page size of approximately seventeen centimetres by twenty-four centimetres, with one thousand six hundred and ninety pages in the body of the dictionary, and with over one hundred and sixty-two thousand references. The *Concise* has a page size of approximately thirteen by twenty-one centimetres, with one thousand three hundred and seventy-nine pages in the body of the dictionary, and over ninety-six thousand references. The *Pocket* has a page size of approximately ten-and-a-half by eighteen centimetres, with nine hundred and ninety-two pages in the body of the dictionary, and seventy thousand references. These measures of comparison are summarised in the table below:

[1]	FORMAT	PAGES	REFERENCES
CED	17×24 cm	1690	> 162000
Concise	13×21 cm	1379	> 96000
Pocket	10.5×18 cm	992	70000

An explanation is needed of the term **references**, which has to do with the way in which the entries in the dictionaries are counted. This is based, according to Mr William McLeod of Collins (personal communication), on a standard American system of entry counting, according to which the following count as 'references': each headword (printed in bold); all run-ons and subentries (including derived words and idioms), also printed in bold; inflected forms that are actually shown in the articles (e.g. irregular past tenses like *saw*); all changes of word-class (part-of-speech) within a headword (e.g. *skin* noun and *skin* verb count as two references). The number of references claimed by Collins are arrived at on a statistical sampling basis. The American system was devised to provide a measure of comparability between dictionaries, at the insistence of the American Government Purchasing Department.

The *CED* is the progenitor of the Collins range, with a first publication date of 1979 (Second Edition 1986). Next to be published was the *Pocket* dictionary in 1981, which refers in its Foreword to the *CED* as 'its larger brother'. The reduction in size has been achieved by 'concentrating on a judicious selection of general and special vocabulary that excludes rarer and highly technical words and meanings . . . and by omitting proper names'. The *Concise* dictionary followed a year later in 1982 and in the Foreword refers to the *CED* as 'its parent'. The Foreword explains 'the strategy by which the content of the *Collins English Dictionary* has been reduced to the practical and workmanlike compass of this Concise version'. It consisted of: the omission of some rare, obsolete and very technical words; the 'judicious' omission of some rare, obsolete and technical senses of words; and by 'skilful' merging of related senses and reducing the length of individual definitions. However, the Foreword also notes that the *Concise* contains some lexemes not included in its larger parent, because they had been noted since the publication of *CED*.

Let us now take our comparison to the content of the dictionary, and first of all to the number of headwords. By way of random sample we will consider the lexemes beginning with *fri-*. Excluding proper names of persons and places (contained only in *CED*) we find that *CED* has seventy-four headwords beginning with *fri-*, the *Concise* has fifty, and the *Pocket* has forty. The headwords contained in *CED*, but not in the *Concise* are as follows:

> [2] friar bird   Friar minor   friar's lantern   fribble
> friction clutch   friction match   friction tape
> Friesian   frig   frigging   frightfully   frigorific
> Frimaire   fringed orchis   fringe tree   fringilline
> frippet   Fris.   frise   frisette   friseur   frisket
> frit fly   frivol   frizette

The *Concise* has one headword not contained in *CED*: *Friends of the Earth* – giving a net figure of twenty-four additional headwords in *CED* over the *Concise*. Of the twenty-five items listed in [2] about half could be considered technical or specialised, three are alternative terms (*friar's lantern*) or spellings (*Friesian, frizette*) cross-referenced to other headwords in the dictionary, three are marked 'taboo slang' (*frig*) or 'informal' (*frivol*), one is an American English word (*friction tape*), one is obsolete (*frigorific*), and one is contained in the *Concise* as a run-on item (*frightfully*). In the case of *fribble* and *frippet*, the editors must have decided that they are unusual or rare lexemes, though they are not marked as such in *CED*.

The headwords contained in the *Concise* but not in the *Pocket* are as follows:

> [3] fricandeau   Friend   friend at court   Friends of
> the Earth   Friesian   frieze2   frigate bird   frijol
> frilled lizard   fringing reef   frisson   frit   frittilary

The *Pocket* has three headwords not contained in the *Concise: Fridays, friendship* (included as a run-on under *friend* in the *Concise*), and *frith* (defined as a variant of *firth*, under which headword it is noted in the *Concise*). *Friend*

(='Quaker') is included in the *Pocket* as a sense of *friend*, and *friend at court* is a run-on item under the same headword. All the other lexemes in [3] could be regarded as technical or specialised in some way. It is particularly striking how more species of animals, birds, plants, etc. are included the larger the dictionary.

Now let us narrow our focus even more and compare a sample entry from the three dictionaries in the Collins range. We will consider the entries for the lexeme *friend*, given at [4], [5] and [6] for the *CED*, the *Concise* and the *Pocket* respectively.

[4] **friend** (frɛnd) *n.* **1.** a person known well to another and regarded with liking, affection, and loyalty; an intimate. **2.** an acquaintance or associate. **3.** an ally in a fight or cause; supporter. **4.** a fellow member of a party, society, etc. **5.** a patron or supporter: *a friend of the opera.* **6. be friends (with).** to be friendly (with). **7. make friends (with).** to become friendly (with). ~ *vb.* **8.** (*tr.*) an archaic word for **befriend**. [Old English *frēond*; related to Old Saxon *friund*, Old Norse *frǣndi*, Gothic *frijōnds*, Old High German *friunt*] – **friend+less** *adj.* – **friend+less+ness** *n.* – **friend+ship** *n.*

[5] **friend** (frɛnd) *n.* **1.** a person known well to another and regarded with liking, affection, and loyalty. **2.** an acquaintance or associate. **3.** an ally in a fight or cause. **4.** a fellow member of a party, society, etc. **5.** a patron or supporter. **6. be friends (with).** to be friendly (with). **7. make friends (with).** to become friendly (with). ~ *vb.* **8.** (*tr.*) an archaic word for **befriend**.[OE *frēond*] – **'friendless** *adj.* – **'friendship** *n.*

[6] **friend** (frend) *n.* [OE frēond] **1.** a person whom one knows well and is fond of. **2.** an ally, supporter or sympathizer. **3.** [F-] a member of the Society of Friends; Quaker. – **friend at court** an influential acquaintance who can promote one's interests. – **friend'less** *adj.*

Compare [4], [5] and [6]. What has been done to condense the entry for *friend* in the *Concise* and *Pocket* editions?

The *Concise* entry [5] contains the same eight senses as the larger *CED*, though the definitions have been trimmed slightly, e.g. the synonyms 'an intimate' and 'supporter' in senses **1** and **3** have been omitted, as has the illustrative example in sense **5**. One of the derived words (*friendlessness*) has been omitted. But the most striking reduction comes in the amount of etymological information included: the *Concise* gives only the Old English form from which *friend* originates and omits all the cognate words in related languages. We might also note that the *Concise* uses the abbreviation 'OE', whereas *CED* gives this in full. Essentially, though, the entry in the *Concise* can be viewed as a trimmed version of the entry in its larger parent dictionary.

More than a trim, however, has been applied to obtain the entry for *friend* in the *Pocket* dictionary [6]. Here *friend* has only two senses: sense **1** corresponds to the first sense in *CED* and the *Concise*, though it is worded differently. Sense **2** represents a collapsing of the second to fifth senses of the other dictionaries, in a much truncated form. The idiomatic expressions 'be friends (with)' and 'make friends (with)' are omitted altogether, as is the (archaic) verb sense. A third sense is included in the *Pocket* edition (*Friend* = 'Quaker'), which constitutes a separate headword in the other two dictionaries, as does the expression 'friend at court'. The *Pocket* lists only one derived lexeme, the adjective *friendless*. Like the *Concise*, the *Pocket* has the minimum of etymological information, and it takes up an early position in the article. You may also note that the *Pocket* does not use the IPA for indicating pronunciation: in the case of *friend* this can be deduced from the use of the symbol *e* instead of *ɛ*. The *Pocket* does not bear the same degree of family likeness to the *CED* that the *Concise* does; this may be because of the more severe condensation required or because the family relationship is different – the *Pocket* is not an offspring of the *CED*, more a minor sibling.

We have considered the differences between the *CED*, the *Concise* and the *Pocket* in terms of their relative sizes and therefore of the amount of information that can be packed into each of them. An alternative perspective would be to view these dictionaries as being aimed at different groups of potential users. The *Concise* version would probably be aimed at general family use and at secondary/high school pupils: as we have seen, it contains essentially the same range of strictly lexicographic information, with a little trimming, as the larger *CED*. The *CED* represents the comprehensive record of the language for the serious dictionary user, for college students and for libraries. The *Pocket*, on the other hand, claims to be 'an ideal reference book for home, school, and office' (dust jacket blurb), essentially then for those who need a dictionary for the two commonest uses (see Chapter 13), i.e. checking spellings and looking up the meanings of 'hard' words, though ironically it is the 'hard' words that tend to be left out in the process of truncation. As a truncated version, it has no pretensions to being a lexicographic record of the language.

Specialist dictionaries

There are two broad groups of what we might call specialist dictionaries: those that provide specialist information, often a more detailed treatment of information given in general-purpose dictionaries; and those which are aimed at a special group of users. Specialist information dictionaries would include those concerned with spelling, pronunciation, etymology, names (of places or people) and special registers or fields (e.g. slang, botany, computing, medicine). Specialist user dictionaries would include dictionaries for children, foreign learners and crossword or word puzzle and games enthusiasts.

Spelling dictionaries are not popular in the English-speaking world, unlike, for instance, in Germany, where the *Duden Rechtschreibung* is the most commonly bought dictionary. A spelling dictionary is basically an alphabetical list, especially of words that cause spelling difficulties, arranged in such a way that users can easily check the spelling of the item they are unsure about. *Cassell's Spelling Dictionary*, for example, lists 'root words' in alphabetical order, and then under each root word the inflectional and derived forms, presumably on the argument that most spelling problems are at morpheme boundaries (e.g. between a noun root (*story*) and the inflectional suffix for plural (-*(e)s* – *stories*). The entry for *fulfil* reads:

[7] fulfil
fulfill (in italics to show a variant American spelling)
fulfilled
fulfilling
fulfils
fulfilment
fulfillment (variant American spelling)

Pronunciation dictionaries are more commonly found, especially ˙Daniel Jones' *Everyman's English Pronouncing Dictionary*, published in many editions and edited after Jones' death by his former pupil, A. C. Gimson, and since the latter's death by Dr Susan Ramsaran. Jones' dictionary contains approaching sixty thousand words with their pronunciation given in the International Phonetic Alphabet (IPA). The accent that is represented by the IPA transcriptions is the one known as 'Received Pronunciation' or 'RP', the one which, Jones states in his introduction, 'I believe to be very usually heard in everyday speech in the families of Southern English people who have been educated at the public schools'. He acknowledges that only 'a rather small minority' actually use this accent, that it has no particular aesthetic merit, and that it is not to be considered a model – 'I take the view that people should be allowed to speak as they like'. Jones has two reasons for representing this accent in his *English Pronouncing Dictionary*: one is that he believes it to be an accent that is 'readily understood in most parts of the English-speaking world'; and the other is that it 'happens to be the only type of English pronunciation about which I am in a position to obtain full and accurate information'. Jones' caveats have largely gone unnoticed; his dictionary – like dictionaries in general (see Chapter 13, p. 200) – has achieved an authority it was never intended to have, and general-purpose dictionaries (British ones at least) have followed him in using RP as the pronunciation to be represented. As with a spelling dictionary, a pronouncing dictionary is able to present more detailed information than a general-purpose dictionary has space for, e.g. variant pronunciations and the pronunciation of derived words, as well as the pronunciation of common or problematic proper nouns (e.g. *Fleance, Islay*). Consider the following entries from Jones:

[8] **plausibility** ˌplɔːzə'biliti [-zi'b-, -lət-]
   plausib/le -ly, -leness 'plɔːzəb/l [-zib-], -li, -lnis

Another kind of information that is treated in a more expanded from in a specialist dictionary is etymology. We saw that this is given in rather summary form, at least in concise and smaller dictionaries. An etymological dictionary aims to give for each word included the earliest recorded date, together with its previous history and its development in form and meaning. Compare the entries for the etymology of *prosecute* as given in the *Oxford Dictionary of English Etymology* at [9], and in the *Longman Concise English Dictionary* at [10].

[9] **prosecute** prɔ'sikjūt follow up, go on with XV; carry on; institute legal proceedings against XVI. f. *prōsecūt-*, pp stem of L. *prō-sequī* pursue, accompany, f. *prō* PRO-+ *sequī* follow (see SEQUENCE). So **prosecu**.TION. XVI. - OF or late L. Cf PURSUE.

[10] [ME *prosecuten*, fr L *prosecutus*, pp of *prosequi* to pursue – more at PURSUE].

The *Oxford* specialist dictionary entry is clearly more detailed than the *Longman* general-purpose dictionary information. In particular, the *Oxford* entry shows that the two senses of *prosecute* entered the language at different times: 'follow up, etc.' in the fifteenth century (XV), and 'institute legal proceedings' in the sixteenth (XVI). Similarly, the *Oxford* entry is more explicit in the derivation from Latin *prōsequī*, including the cross-reference to *sequence*; and the origin of the noun *prosecution* (sixteenth century) is also given.

The specialist dictionaries we have considered so far have been providing information in a more detailed and explicit form that is given for each entry in at least the larger general-purpose dictionaries. Another kind of specialist dictionary offers a particular selection of the vocabulary of the language. There are, for instance, dictionaries of names – of people and places – which are mostly concerned with the origins of names and their original meanings, such as the *Concise Oxford Dictionary of English Place Names*. There are dictionaries of the special register of slang, such as Eric

Partridge's *Dictionary of Slang and Unconventional English*, which claims to be 'a full and documented account of English slang over four centuries' and includes colloquialisms and catchphrases, fossilised jokes and puns, general nicknames, vulgarisms, etc.

Similarly, the foreign words and phrases, which are often relegated to an appendix in general-purpose dictionaries, are collected together into a dictionary, such as Alan Bliss' *Dictionary of Foreign Words and Phrases*, which contains more than five thousand such items. Bliss has a long and detailed introduction on the nature and origins of foreign words and phrases in English, and he identifies eleven types of foreign expression, including: those for which no reasonable equivalent exists in English, e.g. *chic, esprit de corps*; those which 'display a markedly felicitous turn of phrase', e.g. *ich kann nicht anders*; those which are part of the technical vocabulary of some profession, e.g. *allegro, fresco*; those which refer to foreign institutions or things, e.g. *chateau, dacha*; and those which are used to convey local colour, e.g. *gendarme, Rathaus*.

For the purpose of providing fuller coverage than is possible within the scope of most general-purpose dictionaries there are the many specialist field dictionaries on almost every conceivable subject from anatomy to zoology, and computing to sociology. Such dictionaries, often intended for beginning students (or lay people) in the particular field of study, offer both a greater range of the technical vocabulary than do general-purpose dictionaries and more detailed definition and cross-referencing to other items in the field. They also tend to go beyond definition into a discussion of the concepts being defined. There is some advantage to having all the vocabulary of a particular subject collected together in one book, rather than scattered through the many pages of a general dictionary. Compare the entries for *cursor* from *A Dictionary of Data Processing and Computer Terms* at [11] and the corresponding sense in the entry for *cursor* in the *New Collins Concise English Dictionary* at [12].

> [11] A cursor is a moving spot on a video screen which
> indicates the next position for entering data on the
> screen. Sometimes a winking cursor is used,
> which is useful for drawing attention to a specific

element of data or to indicate the part of the screen in use when entering data or instructions. The keyboard of a microcomputer has cursor control keys for positioning the cursor at particular points on the screen when editing.

[12] a moveable point of light, etc., that identifies a specific position on a visual display unit.

The *Collins Concise* definition is in fact a very good one; other dictionaries do not necessarily give a 'computer' sense at all. But the specialist dictionary entry, which contains only the computer definition and consists of nothing but definition, is more detailed and written in a more explanatory style: the definition proper (the first sentence) is followed by two sentences which expand and explain the definition. In this way, they are sometimes more like encyclopaedias than dictionaries.

The second broad group of specialist dictionaries that we identified were those directed at particular groups of users. An obvious type of dictionary in this category is that intended for foreign learners: we shall consider this type in detail in the next chapter. Another group of users that has special dictionaries aimed at them are children and young people. *The Oxford Children's Dictionary*, for example, is intended 'for young readers who want a real dictionary which is helpfully arranged and easy to understand'; it contains over eleven thousand words which children 'are likely to meet and want to use for themselves'. Aimed at 'older children and young adults' is the *Longman New Generation Dictionary*. Again the aim is to produce for this group of users a real dictionary, 'a work . . . from which they will obtain the real satisfaction that a competent user of traditional reference books enjoys'. The users have been taken into account by how the material is presented on the page (a larger number of short entries rather than fewer longer entries, to ease accessibility), by the usefulness of the lexemes included in the dictionary to that particular age-group (school textbooks were scanned for items, and computer editing ensured adequate coverage of all subject specialisms), and by controlling the vocabulary used in the definitions so that 'the definitions are always written using simpler terms than the words they describe'.

Now compare the entries for *tangle* in the *Oxford Children's Dictionary* at [13], the *Longman New Generation Dictionary* at [14], and the *Collins Pocket English Dictionary* at [15].

[13] **tangle** (tangles, tangling, tangled) **1** To become confused and muddled *Tangled string.* **2** To make something into a confused muddle.

[14] **tangle**[1] *v* **-gled, -gling** to make or become a confused mass of disordered and twisted threads **tangle**[2] *n* **1** a confused mass of hair, thread, string, etc. **2** a confused disordered state **tangle with** *v prep esp. spoken* to quarrel, argue, or fight with (someone)

[15] **tangle** (taŋ'g'l) *n.* [< ?] **1.** an intertwisted, confused mass, as of string, branches, etc. **2.** a jumbled, confused condition ~ *vi.* **-gled, -gling 1.** to become tangled **2.** [colloq.] to quarrel or fight. ~ *vt.* **1.** to make a confused muddle of; intertwist **2.** to catch as in a net or snare. **tan'gly** *adj.*

You will notice immediately that the entries aimed at the younger generation in [13] and [14] are easier to assimilate. This is in part because they omit certain features of the adult dictionary, e.g. pronunciation, etymology; and in the *Oxford Children's* the part-of-speech label as well – indeed only the verb is entered. This dictionary includes the inflected forms in full, to aid spelling, and otherwise contains just simple definitions, the first of which is provided with an example. The *Longman New Generation* lives up to its claim of many entries, separating verb and noun uses, and including the prepositional verb *tangle with* as a separate entry. The definitions of this dictionary are not noticeably simpler than those of the *Collins Pocket* [15], but all dictionaries aim after all to define a lexeme with words simpler than itself. The main feature of children's dictionaries would appear to be the less cluttered presentation, enabling the information to be retrieved more easily.

These dictionaries, like those for foreign learners, are specialist in their target users, though in essence they are general-purpose dictionaries. There are, though, finally, some dictionaries which are aimed at a particular group of users and are specialised in their contents. Into this category would fall *The Dictionary of Anagrams*, which is intended for crossword compilers (of which the author, S. C. Hunter, is one) and solvers of crossword puzzles. Hunter's dictionary has around twenty thousand entries, such as the following:

[16] TROUNCES Cornutes, Construe, Counters, Recounts.

Another such dictionary is *The Penguin Rhyming Dictionary* by Rosalind Fergusson, which is presumably aimed at versifiers. It contains a series of rhyming lists and an alphabetical index. If you want to find out, for example, the possible words rhyming with *juggernaut*, you first look it up in the index, and you are then referred to a rhyming list:

[17] Agincourt, Argonaut, juggernaut, cosmonaut, aquanaut, aeronaut, astronaut, reimport, davenport, re-export, overwrought, ultrashort (having very short wavelength), aforethought, afterthought, worrywort (habitual worrier).

'Rare' words receive an explanation in brackets.

The lexicographer and the market-place

It is an interesting question, which we will just raise without fully discussing, whether the demands of the marketplace are in conflict with the integrity of what dictionary editors consider to be good lexicographical practice (compare Chapter 15). The answer probably depends on what kind of publication you consider a dictionary to be: on the one hand it may be a scholarly record and description of the vocabulary of a language, of which – from the historical perspective at least – the *Oxford English Dictionary* is the prime example; on the other it may be a reference book about words designed to meet the specific needs of an identifiable group of users. Perhaps the conflict arises when these two aims of lexicography are confused.

However, it may well be the case that radical innovation in dictionary design, layout and content – proposed as a result of developments in lexicological theory (see Chapter 16, p. 248) – is not possible because the public, and the publishers, have a fixed idea of what a dictionary should look like, deriving from a tradition developed over centuries. Say, for example, that a lexicographer decided that definitions as we know them were not the best or most appropriate way in which to describe the meaning of words, but that words were best defined by a list of carefully chosen examples. Would this lexicographer ever get such a dictionary published?

Exercises

1. Compare the following entries for *pea* and describe the differences.
   (*CED*) **pea** (pi:) *n.* **1.** an annual climbing papilionaceous plant, *Pisum sativum*, with small white flowers and long green pods containing edible green seeds: cultivated in temperate regions. **2.a.** the seed of this plant, eaten as a vegetable. **b.** (*as modifier*) *pea soup.* **3.** any of several other leguminous plants, such as the sweet pea, chickpea, and cowpea. **4. the pea** *Austral. informal* the favourite to succeed. [C17; from PEASE (incorrectly assumed to be a plural)] – '**pea + like** *adj.*
   (*Collins Concise*) **pea** (pi:) *n.* **1.** an annual climbing plant with small white flowers and long green pods containing edible green seeds: cultivated in temperate regions. **2.** the seed of this plant, eaten as a vegetable. **3.** any of several other leguminous plants, such as the sweet pea. [C17: < PEASE (incorrectly assumed to be a pl.)].
   (*Collins Pocket*) **pea** (pe) *n.* pl. **peas,** archaic **pease** [< ME *pese,* a pea, taken as pl. ult. < Gr. *pison*] **1.** a climbing plant with green seedpods. **2.** its small, round seed, eaten as a vegetable. ~**as like as two peas (in a pod)** exactly alike.
2. Compare the following etymologies for *juggernaut* and describe the differences.
   (*Collins Pocket English Dictionary*): 'after Hindi *Jagan-*

nath, an incarnation of the Hindu god Vishnu'
(*Longman Concise English Dictionary*): 'Hindi *Jagannāth*,
title of Vishnu, lit., lord of the world; fr a former belief
that devotees of Vishnu threw themselves beneath the
wheels of a cart bearing his image in procession'
(*Oxford Dictionary of English Etymology*): 'title of
Krishna, avatar of Vishnu; idol of this carried in an
enormous car, under which (it was once said) devotees
threw themselves. XVII; also fig. – Hindi *Jagannāth* –
Skr *Jagannātha*, f. *jagat-* world + *nāthās* lord, protector'

3. Compare the following entries for *flannel*, and describe
the differences.

(*Oxford Children's Dictionary*) **flannel** (flannels) **1.** A
kind of soft cloth **2.** 'Flannels', trousers made of this
3. *A face-flannel*, a piece of cloth used for washing
oneself.
(*Longman New Generation Dictionary*) **flannel** *n* **1.** a kind
of smooth loosely-woven woollen cloth with a slightly
furry surface **2.** a piece of cloth used for washing
oneself; facecloth **3.** *esp. spoken* meaningless though
attractive words
(*Collins Pocket Dictionary*) **flannel** (flan'l) *n.* [< W.
'gwlan', wool] **1.** a soft, loosely woven woollen cloth.
2. a small cloth, usually of towelling, used in washing.
3. (*pl.*) trousers, etc. made of flannel. **4.** [colloq.] evasive
talk; flattery ~ *vt.* **-nelled, -nelling** to wash with a
flannel. **2.** (*colloq.*) to flatter -**flan'nelly** *adj.*

4. Look in the dictionary section of (a) a library, and (b) a
good bookshop. What other types of dictionary do you
find apart from those mentioned in this chapter? How
would you classify them? Look out, for example, for
dictionaries of new words, synonyms, usage, abbreviations and acronyms, etc.

5. Take the entry for *friend* at [5] above (from the *Collins
Concise*) and rewrite it for a children's dictionary. Then
compare your attempt with the entry in an actual children's dictionary.

CHAPTER 12

Especially for the Learner

We passed over one type of specialist English dictionary in the previous chapter: that for the foreign learner of English. Like children's dictionaries, those for the foreign learner are essentially general-purpose dictionaries, but tailored to the needs of a specific group of users. We may well wonder whether there is any justification for such a dictionary. Aren't foreign learners served well enough by bilingual dictionaries? Or if not by them alone, then by the range of ordinary monolingual dictionaries in addition?

What is different about learners' dictionaries?

Compare the entries for the verb *inform* from the bilingual *Collins/Klett English-German Dictionary* at [1], the *Longman Concise English Dictionary* (*LCED*) at [2], and the learner's *Longman Dictionary of Contemporary English* (*LDOCE*) at [3]. Do the differences reflect the different aims of the three dictionaries?

[1] **inform** [in'fɔːm] **I** *vt person* benachrichtigen, informieren (*about* über + *acc.*); unterrichten.
to __ sb of sth jdn von etw unterrichten, jdn über etw informieren; **I am pleased to __ you that** . . . ich freue mich, Ihnen mitteilen zu können, daß . . .; **to __ the police** die Polizei verständigen *or* benachrichtigen *or* informieren; **to keep sb/oneself __ ed** jdn/sich auf dem laufenden halten (*of* über + *acc.*) . . . [*etc.*]

[2] **inform**/in'fawm/*vt* **1** to impart an essential quality of character to **2** to communicate knowledge to ~

vi **1** to give information or knowledge **2** to act as informer *against* or *on* [ME *informen*, fr MF *enformer*, fr L *informare* to give shape to, fr *in—* + *forma* form] **– informant** *n*

[3] **in.form** /in'fɔ:m‖- ɔ:rm/ *v* [T (**of, about**)] *usu. fml* to give information or knowledge to; tell: *I wasn't informed of the decision until too late.| Why wasn't I informed?* [+*obj*+*(that)*] *I informed him that I would not be able to attend.* [+ *obj* + *wh-*] *Could you please inform me how to go about contacting a lawyer?* – see SAY(USAGE)

I have included only about two-thirds of the entry from the *Collins/Klett* bilingual dictionary at [1], so that this entry is in fact considerably longer than the other two. Neither the bilingual dictionary [1] nor the learner's dictionary [3] contain etymological information, presumably because it is considered to be of no relevance or help to a learner of the language – an assumption that is not completely unchallenged. It has been argued that knowledge of etymology may help some learners to understand and retain new vocabulary items (see A. Ellegård, 1978, pp. 225–44). The first sense of *inform* in *LCED* [2] is not mentioned in the other two entries: it is the older sense of *inform* and is put first in *LCED* for this reason, even though it is a rarely used sense in current English; learners would hardly need to be aware of it. The intransitive (vi) senses of the *LCED* entry are not dealt with in the other two dictionaries either, though they both have separate entries for *inform against/on*. The bilingual and the learner's dictionaries have selected the most important and central meaning of *inform*, and they have ignored those not likely to be met by the learner.

Beyond the question of selection, the striking feature of the *Collins/Klett* and the *LDOCE* entries is the attention that they pay to how the word is used in the language, and here the differences between these two dictionaries is most marked. The entry for *inform* in the English-German bilingual dictionary is constructed with translation into German in mind. German glosses constitute the definitions, and the distinctions of meaning that are made reflect the different translations of those meanings in German. So, in the expression 'inform someone of something', *unterrichten* or *informieren* are used in German; but in 'to inform the police' *verständigen* or *benachrichtigen* are used, in addition to *informieren*. Two particular sets of learners are the target users of this dictionary: English speakers learning German, and German speakers learning English. In the case of *LDOCE*, on the other hand, no particular mother tongue is assumed: here the definition is kept deliberately simple, but very detailed syntactic information is given in the codes 'T (of, about)', '[+obj+(that)]', etc; and there is an example to illustrate each of the syntactic constructions – and the meaning.

The three types of dictionary are, then, fulfilling different aims and purposes: specifically, the monolingual learner's dictionary is providing detailed information about usage which is irrelevant to native speakers, and which may be contained in a bilingual dictionary, but not in the same systematised fashion. Is a dictionary with such information justified? What would foreign learners use such a dictionary for? Language learners, like all users of language, employ language in two functions: decoding (i.e. listening, reading), and encoding (i.e. speaking, writing). What is demanded of the dictionary is different in the two functions. In decoding – and as far as dictionary use is concerned it is the reading function that is particularly relevant – the learner needs a means of interpreting lexemes in context: the learner's dictionary, as indeed bilingual dictionaries, need to define clearly all the different senses of a lexeme and provide where appropriate register and field labels. In encoding, and it is mainly writing which is in focus here, the learner needs quite different information. The choice of lexeme has presumably been made; the definitions and register/field labels provide a check that the lexeme is appropriate to the context in which the learner wishes to use it; what the learner needs above all is accurate and detailed grammatical information so that correct and natural sentences can be encoded. It is in this last requirement that learners' dictionaries score significantly over monolingual native-speaker dictionaries and over many bilingual dictionaries. What the learner also needs in order to encode natural sentences is appropriate collocational information. More attention is now being given to providing this information in specialist dictionaries catering for the foreign learner.

It may also be considered that beyond a certain stage in language learning, the use of a monolingual learner's dictionary rather than a bilingual dictionary enhances the learning itself. Learners use the language being learned to advance their learning of that language. Indeed in settings where a group of learners has several mother tongues this may be the only type of dictionary that it is practicable to use. Learners' dictionaries, therefore, like children's dictionaries, need to take account of the limited linguistic resources of their users. Learner's dictionaries come in a variety of sizes aimed at learners at different stages. They are perhaps most well established for learners at the advanced stage, and our attention will be focused on the three leading dictionaries for this stage: the *Oxford Advanced Learner's Dictionary of Current English*, the *Longman Dictionary of Contemporary English*, and the *Collins Cobuild English Language Dictionary*. We shall abbreviate these *OALD*, *LDOCE* and *COBUILD* respectively.

What's in a learner's dictionary?

Let us now look at the kinds of information contained in an advanced learner's dictionary and at how these differ from what is found in a comparable monolingual concise dictionary. Let us make another comparison of entries. Compare the entry for the noun *initiative* in *LCED* at [4] and in *LDOCE* at [5]. How do they differ?

[4] **initiative** /i'nish(y)ətiv/*n* **1** a first step, esp- in the attainment of an end or goal **2** energy or resourcefulness displayed in initiation of action **3** a procedure enabling voters to propose a law by petition – compare REFERENDUM – **on one's own initiative** without being prompted; independently of outside influence or control

[5] **i.ni.tia.tive** /ɪ'nɪʃətɪv/ *n* **1** [U] *apprec* the ability to make decisions and take action without asking for the help or advice of others: *I wish my son would*

show a bit more initiative.|Don't keep asking me for advice; **use your (own) initiative.** 2 [C] the first movement or action which starts something happening: He **took the initiative** in organizing a party after his brother's wedding.|The government is making some fresh initiatives to try to resolve the dispute. 3 [the + S] the position of being able to take action or influence events: Because of a stupid mistake, we lost the initiative in the negotiations; the other side has the initiative now. 4 **on one's own initiative** (done) according to one's own plan and without help; not suggested by someone else

---

Both entries indicate the pronunciation of *initiative*, but in different notations. *LCED* assumes the user to be ignorant of the International Phonetic Alphabet symbols and uses a home-grown notation based on the Roman alphabet plus the 'schwa' symbol /ə/. *LDOCE* however uses the IPA notation, since this is widely used in teaching and learning English as a foreign language, for which it was in large part initially developed. Both entries indicate the word–class (part-of-speech). And both entries have the same first two sense divisions, though in reverse order. But the third sense of the *LCED* entry is not included in *LDOCE*; presumably the editors thought it not common enough to be worth including, though the *OALD* has it in its entry for *initiative*. And *LDOCE* has a sense (3) which is not in *LCED*, where *initiative* occurs with the definite article. One immediately obvious difference in the entries is that the first two senses in *LDOCE* are marked '[U]' and '[C]' respectively. These symbols stand for 'uncountable' and 'countable'. So, *initiative* in its first sense is uncountable (i.e. it is regarded as an indivisible mass), whereas in its second sense it is countable (i.e. it is regarded as an individualised thing of which there may be more than one). We will just note this additional information given by the learner's dictionary at this point and discuss its significance later on. You will have noticed a significant difference in the wording of the definitions: the words used in the *LDOCE* definitions are markedly simpler than those in the definitions of the Concise. And finally, the *LDOCE* definitions are amply illustrated with examples, which are lacking in the *LCED* entry, though this dictionary does include examples on occasions. Now we will look in more detail at these and other points.

Pronunciation and syntax

We noted above that pronunciation in learners' dictionaries is indicated by means of the widely used alphabet of the International Phonetic Association. This is also used in some native-speaker dictionaries (e.g. *Collins, Pocket Oxford*), but learners' dictionaries in addition indicate where 'General American' pronunciation differs from the British (RP) pronunciation represented. If you look back to the *LDOCE* entry for *inform* at [3] you will note an alternative (US) pronunciation (after the double vertical line) indicated for the end of the word. The indication of American pronunciation acknowledges that learners' models of pronunciation (e.g. teachers, recorded course materials) may be American rather than British. Learners' dictionaries likewise show alternative American spellings, but so do most up-to-date native-speaker dictionaries. Both kinds of dictionary also indicate irregular inflectional forms of lexemes (discussed in Chapter 10, p. 148).

Let us turn now to the syntactic information given in learners' dictionaries, which is far more detailed than that found in native-speaker dictionaries. In the latter, syntactic information is limited to the word–class label and the division of verb senses into transitive and intransitive. In learners' dictionaries, the tradition of giving the word–class of a lexeme is continued: this, as we saw in Chapter 10 (p. 145) gives the most general of information about the syntactic operation of lexemes. More detailed syntactic information is usually given by means of codes.

In both *OALD* and *LDOCE* most nouns, or senses of nouns, are coded with either 'C' or 'U', and correspondingly 'N COUNT' or 'N UNCOUNT' in *COBUILD*. See, for example, the *LDOCE* entry for *initiative* at [5] above, and the relevant extracts from the *OALD* entry for *difficulty* at [6] and from the *COBUILD* entry for *newspaper* at [7] following.

[6] **difficulty** . . . *n* **1** [U] the state or quality of being difficult . . . **2** [C] (pl -ties) sth difficult, hard to do or understand . . .

[7] **newspaper** . . **newspapers**. **1** A **newspaper** is **1.1** a publication N COUNT consisting of a number of large sheets of folded paper, on which news, advertisements, and other information is printed . . . **1.2** an N COUNT organization that produces a newspaper.

2 Newspaper consists of pieces N UNCOUNT of old newspapers especially when they are being used for another purpose such as wrapping things up.

The information provided by these codes relates not only to whether the noun may be used in the plural (countable) or not (uncountable), but also to the possible determiners (e.g. *a, the*) and quantifiers (e.g. *many, some, few*) that may occur with the noun. For example, a countable noun (or the countable sense of a noun) may occur with *a* in the singular or with *the* in the singular or plural and may occur with quantifiers like *many, few, several* and with *some* in the plural. Uncountable (or mass) nouns, on the other hand, may occur with *the*, but not with *a*, and may occur with quantifiers *some* (in the singular), *little, much*, but not with *many, several*, etc. These syntactic possibilities in the noun phrase represent important information for the foreign learner, since the countability of nouns in other languages does not always correspond with English usage; e.g. *information* is uncountable in English, but countable in some other languages.

The most detailed and extensive syntactic information in learners' dictionaries is that given for verbs. This is not surprising, since verb syntax is essentially the syntax of the clause, and it is where there are probably more differences between languages. The verb lexeme in a clause determines the potential occurrence of the other elements in the clause, e.g. subject, objects, complements. Consider, for example, the following clauses containing the verb *steer*. What can you conclude from them about the syntax of *steer*?

[8] I'm steering. (You look out for the signposts.)
[9] They steered for port.
[10] This truck doesn't steer very well.
[11] The driver steered the truck into the parking place.
[12] The captain steered a course for Liverpool.

In a native-speaker dictionary, *steer* would be marked simply as 'v' or alternatively as 'vi, vt', to indicate that it may be used either without an object (i.e. intransitively, as in [8] to [10]) or with an object (i.e. transitively, as in [11] and [12]). But, as we can see from these examples, there is more to the syntax of *steer* than just the alternation between intransitive and transitive. Besides the subject, which is present in all the examples, *steer* is accompanied by no other element in [8], by an adverbial of direction in [9], by an adverbial of manner in [10], by an object and an adverbial of direction in [11] and [12]. [12] differs from [11] in having a different kind of object (not a vehicle but an abstract notion). We may also note that the subject in [10] does not refer to the person steering, as the subjects do in all the other sentences, but to the vehicle being steered. For foreign learners it is important to be informed of these syntactic restrictions and possibilities, so that they may be able to construct permissible and appropriate sentences in English. Compare now the entries for *steer* from *OALD* at [13], *LDOCE* at [14] and *COBUILD* at [15], and note how the syntactic information is presented.

[13] **steer**² /stɪə(r)/ *vt, vi* [VP6A, 2A, C] direct the course of (a boat, ship, car, etc): ~ *north*; ~ *by the stars*; (with passive force): *a ship that* ~ *s* (= is ~ ed) *well/easily/badly*. ~ ***clear of***, (fig) avoid.

[14] **steer**¹ /ˈstɪəʳ/ *v* 1 [I;T] to make (esp. a boat or road vehicle) go in a particular direction: *She steered with one hand while trying to adjust the rear-view mirror with the other.|He steered the boat carefully between the rocks.|(fig.) I steered the visitors towards the garden.|(fig.) She tried to steer the conversation away from such dangerous topics.|(fig.) steering a bill through Parliament* **2** [I + adv/prep; T + obj + adv/prep] to follow or change to (a particular course), esp. in

a boat: *We turned about and steered (a course) for Port-of-Spain.|to steer a middle course between two extremes* **3** [I + *adv/ prep*] (of a boat or vehicle) to act when one turns its steering wheel; *How does your car steer? Does it take the corners well?* **4 steer clear (of)** *infml* to keep away (from); avoid: *I should steer clear of the fish stew; it's not very nice!*

■ USAGE You can **steer** *ships, cars, lorries,* etc., and also such things as *cycles* and *sledges,* but not *aircraft*; for these, the usual verbs are **fly** and **pilot.** – see also BOAT (USAGE), CAR (USAGE), DRIVE (USAGE)

[15] **steer** /stɪə/, **steers, steering, steered. 1** When you **steer** a car, boat, plane, etc, you operate it so that it goes in the direction   V-ERG OR V that you want. EG *They set off* ⇑ guide *with no idea how to steer a boat . . . He steered the car through the broad entrance . . . The freighter steered out of Santiago Bay that evening.*

2 If you **steer** someone in a   V + O: USU + particular direction, you guide A them there, for example by = propel putting your hand on their arm or back and pushing them very gently. EG *He steered me to a table and sat me down in a chair.*

3 If you **steer** people towards a   V + O: USU + particular course of action or A way of behaving, you change ⇑ guide their behaviour, especially without them noticing, by guiding them into this course of action or way of behaving. EG *The leader had steered the party away from communism . . . He steers the conversation towards more general topics.*

4 If you **steer** a particular   V + O course, you take a particular line = follow of action. EG *The course he steered was perilous . . . The panel finally steered a judicious middle course.*

OALD retains the traditional 'vi, vt' labels, but indicates the syntax of verbs in more detail by means of a series of 'VP' (i.e. verb pattern) codes. There are twenty-five verb patterns given for English, but many have subdivisions, and the total amounts to over fifty. They are based on a scheme for describing verb syntax devised by A. S. Hornby, the original editor of *OALD*, and presented in detail in his *A Guide to Patterns and Usage in English* (1954). 'VP6A' is a pattern composed of 'S(ubject) + vt + noun/pronoun', where the clause may be made passive (there is an identical pattern '6B' where passivisation is not permissible); 'VP2A' is composed of 'S + vi'; and 'VP2C' has 'S + vi + adverbial adjunct'. These patterns do not provide for adverbials which are considered to be optionally occurring. So, our sentences [8] and [9] are instances of 'VP2A', [10] is an example of 'VP2C', while [11] and [12] exemplify 'VP6A'.

LDOCE codes syntactic information for verbs in two ways. It indicates the traditional division between intransitive and transitive by means of the symbols 'I' and 'T'. If a verb may enter a syntactic pattern with more than just an intransitive verb or a transitive verb + noun/pronoun object, then that is indicated at the sense or example to which it refers. In the case of *steer*, sense 2 is coded '[I+adv/prep; T+obj+adv/prep]': this indicates that in this sense, *steer*, when used intransitively, is followed by an obligatory adverbial or prepositional phrase, and when used transitively, it similarly has an adverbial or prepositional phrase in addition to the object. So, our sentence [8] is an instance of sense **1** 'I', sentence [9] an example of sense **2** 'I+adv/prep', sentence [10] an example of sense **3** 'I+adv/prep', sentence [11] of sense **1** 'T', and sentence [12] of sense **2** 'T+obj+adv/prep'.

COBUILD puts all its grammatical information into what is called the 'extra column', to the right of the definitions. The syntax of verbs is indicated by means of formulas like 'V+O', interpreted as 'verb + object'. In the entry for *steer* at [15] the first sense has the syntactic designation 'V-ERG OR V', i.e. ergative verb or verb. The term 'ergative' refers to the alternation between a transitive and an intransitive pattern, in which the object of the transitive pattern appears as the subject of the intransitive pattern: this corresponds to the relation between our sentences [10] and [11]. The simple 'V' by itself indicates an intransitive use of the verb, in which, if there is a correspondence with a transitive (V+O) pattern, the subjects are of the same kind for both. *Steer* enters both ergative and non-ergative intransitive sentences. The formula 'V+O: USU+A' indicates that these transitive uses of the verb usually include an 'adjunct' (or adverbial) in the syntactic pattern. Our sentences [8], [9], [10] and [11] all come under sense **1** with its syntax of 'V-ERG OR V'; sentence [12] corresponds to sense **4**, and *a course for Liverpool* would count as 'O' (i.e. object). Senses **2** and **3** are the figurative meanings of *steer*, corresponding to the final examples under sense **1** of *steer* in the *LDOCE* entry at [14].

The three dictionaries thus employ different methods of coding or abbreviation to represent the syntactic operation of verbs. More recently the trend has been to make grammatical coding in learners' dictionaries less abbreviated and impenetrable: this is apparent from comparing the older *OALD* (1974) with its verb pattern codes and the newer *COBUILD* or *LDOCE*. The new edition of *LDOCE* (1987), it may be noted, has a considerably more understandable and accessible coding system than the first edition of 1978. It must be said, however, that Hornby's verb patterns in the *OALD* (first edition 1948) blazed the trail for the inclusion of detailed syntactic information in learners' dictionaries, which others have then followed. Another welcome development has been the way in which examples are used to illustrate not only the meanings but also the syntactic operation of words. In the *LDOCE* and *COBUILD* entries virtually every syntactic pattern has at least one example sentence to illustrate it, and the examples are not concocted by the lexicographer but taken from genuine language data.

Consider now the following entries from *LDOCE* at [16] and *COBUILD* at [17] for the adjective *criminal*:

[16] **crim.i.nal**[1]/'krɪmənəl/*adj* **1** being a crime: *a criminal offence* (= a serious offence, esp. one that

you could be sent to prison for)|*criminal behaviour*|*tendencies* 2 [A *no comp.*] of crime or its punishment: *A criminal lawyer is a specialist in criminal law.* – compare CIVIL (2) 3 *infml* very wrong: *a criminal waste of money* – ~ly *adv*

[17'] **criminal**/krɪmɪnəˈl/

2 Something that is **criminal** is	ADJ CLASSIF
2.1 connected with crime or with the punishment of crime. EG *He had done nothing criminal . . . It is a criminal offence . . . Scotland has its own criminal law.* ◊	◊ ADJ + ADJ/ADV
criminally. EG . . . *the care of the criminally insane . . . They decided that he was not criminally responsible for what had happened.* **2.2** morally wrong, but not illegal.	ADJ QUALIT
EG *To refuse medical aid would be criminal.* ◊ **criminally**. EG *The pay was criminally poor . . . His staff were criminally underpaid.*	◊ ADV + ADJ/ ADV

Most adjectives occur in two normal positions in English syntax: before nouns, i.e. attributively (*the big house*), and after linking (copula) verbs like *be*, i.e. predicatively (*the house is big*). It is important to know when an adjective is restricted to one or other of these positions. The *LDOCE* entry at [16] indicates that in sense 2 *criminal* is restricted to attributive position ('A'): there is no *the lawyer is criminal corresponding to *the criminal lawyer*. With the 'no comp' abbreviation, *LDOCE* indicates for this sense of *criminal* that it may not inflect for comparative or superlative: there is no *the more/most criminal lawyer*. See the entry for *afraid* at [18] for an adjective restricted to predicative position. While the *COBUILD* entry at [17] indicates neither of the pieces of syntactic information contained in *LDOCE*, it does make another distinction between its two senses of *criminal*. Sense **2.1** is designated in the extra column 'ADJ CLASSIF', while sense **2.2** has the label 'ADJ QUALIT'. Classifying adjectives are non-gradable: they cannot be modified by an intensifying adverb like *very* nor be used in comparative constructions, e.g. with *more* (this in effect includes the information given by 'no comp' in *LDOCE*).

A qualitative adjective on the other hand may be modified by *very* and enter constructions with *more* and the like. Moreover, if a classifying and a qualitative adjective occur together, then the classifying adjective follows the qualitative one (e.g. *a good criminal lawyer, a criminal official practice*).

The other kind of syntactic information required for adjectives relates to adjective complementation, i.e. what categories of element may be required by an adjective in predicative position. Look at the following *LDOCE* entry for the first sense of *afraid*:

> [18] **a.fraid** ə'freɪd/*adj* [F] **1** [(**of, for**)] full of fear; frightened: *There's no need to be afraid.|Don't be afraid of the dog.|He was afraid for his job.* (= afraid that he might lose it) [+ *to- v*] *I was afraid to go out of the house at night.* [+ (*that*)] *They were afraid that the police would catch them.* -see FRIGHTENED (USAGE)

The code 'F' (= 'following') placed before the sense number (**1**) indicates that the adjective is restricted to predicative position in all its senses. For sense **1**, the entry shows that *afraid* may be optionally (hence the brackets) complemented by a prepositional phrase introduced by *of* or *for*. An example is then given for each of the three syntactic possibilities: no complement, *of*-complement, *for*-complement. There then follow two further examples which illustrate additional types of complement with *afraid*. The type of complement is shown by the abbreviations in the square brackets preceding each example: '[+to-v]' indicating a *to*-infinitive clause complement; '[+(that)]' indicating a *that*-clause complement from which the conjunction *that* may be omitted.

Definitions and examples

Let us turn now to the definitions in learners' dictionaries. You will have noticed how much simpler the definitions are for *inform* at [3] and *initiative* at [5] in *LDOCE* than in the corresponding entries from *LCED* at [2] and [4] respectively. The *OALD* similarly claims to provide 'practical definitions in simple English' (on the dust jacket). The editors of *LDOCE* have in fact restricted the words used in their definitions to a list of approximately two thousand specified items, which are listed in the end-matter of the dictionary. Moreover, the definitions have been checked by computer to ensure that only this restricted defining vocabulary has been used. If it has been necessary to go outside of it, then the items show up in the definitions in small capitals, for cross-reference to the appropriate entry elsewhere in the dictionary. Learners' dictionaries therefore take account of the limited vocabulary of their users in the same way that children's dictionaries do. *COBUILD* has taken a more radical approach to the writing of definitions. A definition in *COBUILD* always consists of a complete sentence, so that 'the user of the dictionary is shown the word in natural English' (p. viii) and so that the definition illustrates both the typical grammatical context and the typical use of the word.

Compare now the entries for *prescriptive* from *LCED* at [19], *LDOCE* at [20], *OALD* at [21], and *COBUILD* at [22]. How are the learner's dictionary definitions 'simpler' or more helpful than the native-speaker dictionary definitions?

[19] **1** serving to prescribe **2** established by, founded on, or arising from prescription or long-standing custom **3** authoritarian as regards language use

[20] **1** *tech, sometimes derog* saying how a language ought to be used, rather than simply describing how it is used: *prescriptive grammar* – compare DESCRIPTIVE (2) **2** *fml* saying how something should be done or what someone should do

[21] giving orders or directions; prescribed by custom: *a ~ grammar of the English language*, one telling the reader how he ought to use the language. → descriptive.

[22] Something that is **prescriptive** sets   ADJ QUALIT down rules and states what should= strict and should not happen in certain circumstances; a formal word. EG *He is a man free of prescriptive social*

*norms . . . His account was descriptive
rather than prescriptive.*

LDOCE and *OALD*, which employ a traditional style of definition, would appear from this example to simplify their definitions in different ways. *LDOCE* simplifies by using the restricted defining vocabulary, which tends to lead to rather long and wordy definitions to achieve simplicity. *OALD*, on the other hand, tends to go for terse definitions, with the danger perhaps of oversimplifying the description of meaning. However, in the decoding function, when the dictionary is used as an aid to reading, this broad-brush approach to the description of meaning is perhaps all that is necessary. Both dictionaries contain an example illustrating the linguistic sense of *prescriptive*, perhaps the sense that a foreign learner of English might most readily come across. *OALD* additionally gives an explanatory description of the example, which we may regard as an extension of the definition. This is a feature of this dictionary, which has been adopted by the new (1987) edition of *LDOCE*, though not in this entry, presumably because the definition itself gives adequate explanation.

COBUILD does not single out the linguistic sense of *prescriptive* for special treatment, though it does draw the *descriptive/prescriptive* contrast in one of its examples. The simplification or helpfulness of the *COBUILD* definitions consists in the use of complete sentences in straightforward English, which serve to contextualise the word. Additionally, in the extra column, *COBUILD* indicates, where it is appropriate, synonyms (in this case *strict*), antonyms and superordinate terms related to the lexeme. And *COBUILD* is characterised by an abundance of examples.

As we have seen, the examples in learners' dictionaries constitute a very important part of the entries. In the decoding function of the dictionary they provide a range of typical contexts that aid the distinguishing of the different senses of a lexeme, thus enabling an appropriate interpretation to be made by the user for the particular instance that occasioned the look-up. In the encoding function, the examples serve to illustrate both the possible syntactic environments of (the sense of) a lexeme, and some of its possible collocational or lexical environments, so that the user is enabled to construct grammatically and lexically 'natural' sentences in English. Probably the examples are more important in a learner's dictionary than in any other type of dictionary, because this is often where the user starts in trying to understand the meaning and usage of a lexeme. Ideally, every sense of a lexeme as well as each syntactic code and the main collocations of the lexeme should be illustrated. And the examples should be authentic. Both *LDOCE* (1987) and *COBUILD* recognise the importance of examples. The former claims to have '75000 realistic useful examples based on authentic language from the Longman Citation Corpus', while the latter claims that '96000 examples taken from the COBUILD data base show just how words and phrases are really used' (both quotations from the dust jackets).

More attention has been paid in recent years to tailoring the content and presentation of learners' dictionaries to the needs and abilities of their users than has been devoted to any other kind of dictionary. This no doubt reflects the increase in size and importance of the English-as-a-foreign-language market, but also perhaps the linguistic sophistication of the users and the clear demands articulated by EFL teachers for a practical and usable dictionary. It is to the users of dictionaries that we turn in the next chapter.

Exercises

1. Compare the entries for *rehearse* from *LCED* and *LDOCE* (*Note*: 'I' = intransitive, 'T' = transitive). Give a detailed account of the differences.

    LCED: *vt* **1** to present an account of (again) <~*a familiar story*> **2** to recount in order <*had* ~*d their grievances in a letter to the governor*> **3a** to give a rehearsal of; practice **b** to train or make proficient by rehearsal ~ *vi* to engage in a rehearsal of a play, concert, etc

    LDOCE: *v* **1** [I,T] **a** to practise (a play, concert, etc) in order to prepare for a public performance: *The actors were rehearsing (the play)*

until 2 o'clock in the morning **b** to cause (someone) to do this: *She rehearsed the musicians* **2** [T] *fml* to tell fully (events or a story); RECOUNT

2. What are the syntactic possibilities for the verb *intend*? You may like to check your conclusions with the entry for *intend* in one of the learners' dictionaries.

3. Compare the entries for the noun *cake* from *LDOCE* and *COBUILD*. How do they differ?

LDOCE **cake**[1] /keɪk/ *n* **1** [C;U] (a piece of) a soft food made by baking a sweet mixture of flour, eggs, sugar, etc.: *to bake a cake|a chocolate cake|a birthday cake|Would you like some cake|a slice of cake?* – compare BISCUIT; see also CHRISTMAS CAKE, CUP CAKE, MADEIRA CAKE **2** [C] (*often in comb.*) a flat shaped piece of something, esp. food: *a potato cake|a fishcake|a cake of soap* **3** [the + S] the total amount, esp. of money or goods, that is to be shared among everyone: *The people of the Third World want a bigger slice of the cake.*

COBUILD **cake**/keɪk/, **cakes 1 A cake** is **1.1** a sweet food made by baking a mix- N COUNT ture of flour, eggs, sugar, fat, etc in an oven. Cakes may be large and cut into slices, or they may be small and intended for one person only. EG *She said she would bake a cake for my birthday . . . She cut the cake and gave me a piece.* ▶ used as an un- ▶ N UNCOUNT count noun. EG *. . . a slice of cake . . . I enjoyed sitting down with friends over coffee and cake when the day's work was over.*

1.2 food that has been N COUNT
formed into a flat, round + SUPP
shape, usually before it
is baked or fried. EG
... *fish cakes* ... ...
cakes of pounded rice.
2 A cake of something N COUNT: ALSO
such as soap or wax is a N+of+N
small block of it. EG *He*
was given a pink cake of UNCOUNT
soap, which smelled of = bar
disinfectant.

4. Put yourself in the position of a foreign learner using a
   dictionary. Is there any information which such a learner
   might usefully need that is not contained in the advanced
   learners' dictionaries?

CHAPTER 13

Who Uses a Dictionary for What?

Before we look at what has been discovered about the use of dictionaries – and really very little is known about the patterns of dictionary use – let me ask you to complete a questionnaire about what you use a dictionary for. Fill it in as honestly as you can!

[1] 1. Which dictionary or dictionaries do you own and use?
   2. How often do you refer to a dictionary? (Please tick the answer which corresponds most closely to your frequency of use.)
      – once a week or more
      – once or twice a month
      – less frequently
   3. On what occasions do you use a dictionary? (Please tick as many as are appropriate to your dictionary use.)
      – while reading
      – for writing essays
      – doing crosswords
      – playing word-games
      – general interest
      – other (please specify)
   4. What do you usually use a dictionary for? (Please tick as many as are appropriate to your dictionary use.)
      – looking up meanings
      – checking spelling
      – checking pronunciation
      – checking part-of-speech
      – discovering etymologies

- checking whether a word exists
- other (please specify)
5. Do you think that your dictionary provides you with:
   - all the information that you need? yes/no
   - more information than you need? yes/no
6. How do you think that dictionaries could be improved?

This questionnaire was submitted to fifty students beginning a degree course in English language and literature at Birmingham Polytechnic in Autumn 1986, and to thirty-six students beginning a degree course in Speech and Language Pathology and Therapeutics. You may like to compare your answers to the questionnaire with the results of that survey. We will leave aside the answers to Question 1 for the moment, except to note that two of the (Speech and Language Pathology) students confessed to not owning a dictionary. The second question is directed at the frequency of use of dictionaries. Of the English students, sixty-two per cent claimed to consult a dictionary once a week or more, compared with thirty-three per cent of the Speech and Language Pathology students (for the whole group: fifty per cent). Thirty-two per cent of the English students thought they consulted a dictionary once or twice a month, compared with just over fifty-eight per cent of the Speech and Language Pathology students (whole group: forty-three per cent). This means that six per cent of the English students and a little over eight per cent of the Speech and Language Pathology students (whole group: seven per cent) did not think that they looked at a dictionary more than once a month.

Question 3 is concerned with when people use a dictionary, under what conditions or on what occasions. All the possible occasions listed were ticked by some of the Birmingham Polytechnic students, though the first two, 'while reading' and 'for writing essays', were ticked more often than any of the others, indicating perhaps the educational bias of this sample. Eighty-eight per cent of the English students claimed to use a dictionary while reading, compared with a little more than fifty-five per cent of the Speech and Language Pathology students (whole group: seventy-four per cent); but around eighty-four per cent of both groups of students used a dictionary for writing essays, and a small number of students added under 'other' that they used one for writing letters. It is interesting that the English students use a dictionary almost equally frequently for the decoding function of language (reading) as for the encoding function (writing), while the Speech and Language Pathology students use a dictionary significantly less frequently for decoding than for encoding. The latter students have been schooled more in the natural sciences, and there is perhaps the expectation that difficult words will receive definition and explanation in the textbooks for these subjects. Fewer than fifty per cent of students said that they used a dictionary on the other three specified occasions of use: just above forty-six per cent for crosswords, around forty per cent for word-games, with the scores very similar for the two groups. Slightly more of the English students seemed to consult a dictionary for general interest (forty-six per cent) than of the Speech and Language Pathology students (forty-two per cent).

The fourth question asks what you look for in your dictionary when you use it: the kind of information that you expect to find and want to extract from a dictionary. There must clearly be some correlation between an occasion of use and the information required: word-game and crossword enthusiasts no doubt most often consult a dictionary to check whether a word exists or how it is spelt. If you are using a dictionary while reading, it is likely to be meaning that you are most interested in. The three most frequently ticked uses (by far) were 'looking up meanings', 'checking spellings' and 'checking whether a word exists'. Ninety-six per cent of the English students and eighty-nine per cent of the Speech and Language Pathology students (whole group: ninety-three per cent) said that they used a dictionary for looking up meanings; and for checking spellings the percentages were eighty-eight and ninety-seven respectively (whole group: ninety-two). The third major use, checking whether a word exists, was ticked by seventy-two per cent in both groups. The other three uses suggested in the questionnaire were ticked by very few of the students: between ten and eleven per cent said they used a dictionary for checking pronunciation; six per cent of the English students and under three per cent of the Speech and Language Pathology students used it for checking the part-of-speech of a word; and fourteen per cent and five-and-a-half per cent respectively (whole group: ten-and-a-half per cent) looked up etymological information in the dictionary.

In view of the fact that dictionaries apparently contain a lot of information (pronunciation, part-of-speech, etymology) that they do not make use of, it is surprising that in answer to the second part of Question 5 most of the students did not think that their dictionary contained too much information: fifty-six per cent of the English students and seventy-five per cent of the Speech and Language Pathology students (whole group: sixty-four per cent) answered 'no' to this question; and only ten per cent and nineteen per cent respectively (whole group: fourteen per cent) answered 'yes'. Indeed there seemed to be widespread satisfaction with their dictionaries: to the first part of the question, sixty-six per cent of the English students and over eighty-three per cent of the Speech and Language Pathology students (whole group: seventy-three per cent) answered 'yes'; while just thirty-two per cent and almost seventeen per cent respectively (whole group: twenty-two plus per cent) answered 'no'.

From these results it appears that more of the English students were dissatisfied with their dictionaries than of the Speech and Language Pathology students, and most of the suggestions for improvement under the last question came from the English students. A wide range of suggestions was made for improving dictionaries. Some related to layout and typography, which were thought to be capable of improvement in order to make the dictionary more attractive to consult and the information more readily accessible. Some related to the range of items included: it was felt on the one hand that more colloquialisms, slang, dialect and American usages should be included, and on the other that more scientific and technical terms should be included, along with the expression of concern that the dictionary should be up-to-date. These are, of course, constant concerns of modern lexicographers, and there has been a considerable extension in recent years of the vocabulary included in dictionaries, both from informal and from technical registers of the language.

There were demands that definitions should be longer and more clearly expressed, and that there should be more instances of usage given. It was also felt that more explanation of pronunciation should be given, though most dictionaries do quite well in this area these days. And there was a suggestion that dictionaries should give regional variants of pronunciation. One student in fact proposed that dictionaries should be arranged according to the pronunciation of words, so that they might be more useful for checking spelling. A more interesting suggestion on the arrangement of dictionaries proposed that dictionary and thesaurus should be merged (see the discussion in Chapter 14); and a similar proposal suggested grouping words by subject under each letter of the alphabet. These suggestions, even if from a minority of the students, show that there is still a lot of development left in the craft of lexicography (see Chapter 15).

Professor Randolph Quirk applied a similar though more detailed questionnaire to two hundred and twenty students from a range of disciplines (approximately half from the 'humanities' and half from the 'sciences') in the middle of the first year of their studies at University College London in 1972 (reported in 'The Image of the Dictionary', in *The Linguist and the English Language*, 1974). Quirk's students made somewhat more modest claims about the frequency of their use of a dictionary: of the humanities students, forty-six plus per cent claimed 'weekly' use, thirty-two per cent 'monthly' use, and the remaining twenty-one plus per cent 'infrequent' use; the corresponding percentages for the sciences students were twenty plus, forty-two plus, and thirty-seven respectively. When we come to the information for which dictionaries were consulted by the University College students, we find again that looking up meanings and checking spellings predominate, with little interest in pronunciation, parts-of-speech and etymology. Quirk comments that 'some of the dictionary features which seem of particular centrality to lexicographers are decidedly peripheral to the ordinary user' (p. 154). We might note that the 'ordinary user' referred to here is undergoing a course of higher education and must therefore count among the most educated section of the population.

Use of learners' dictionaries

Our survey of dictionary use has so far been restricted to UK students. There appears to be no published research on the use of dictionaries among the population at large, though publishers sometimes hint that they have undertaken such research. For example, the Preface to the *Heinemann English Dictionary* states: 'Research has shown that most schoolchildren and many adults are deterred from making full use of the range of information which the majority of dictionaries provide.' Presumably publishers are not willing to make known the results of consumer research which they have commissioned, lest they be made use of by rival publishers who have not gone to the expense of undertaking their own.

One other set of English dictionary users that has been investigated, however, are users of advanced learners' dictionaries, such as those we discussed in Chapter 12, p. 177. We noted that the particular feature of these dictionaries, by comparison with monolingual dictionaries for the native speaker, was the wealth of syntactic and usage information, intended to enable learners to encode natural and appropriate sentences in English. The use of these dictionaries was investigated by Henri Béjoint, who submitted a questionnaire to one hundred and twenty-two of his students of English at the University of Lyon, France (reported in 'The Foreign Student's Use of Monolingual English Dictionaries', 1981). Ninety-six per cent of the students possessed a monolingual English dictionary, in most cases one of the two advanced learners' dictionaries discussed in the previous chapter. Over ninety per cent of the students claimed to use the dictionary at least once a week, with forty per cent claiming daily use.

Béjoint asked his students to rank-order seven kinds of information that they might look for in a dictionary; he then computed how many students ranked the kinds of information in the first three places. Top of the list came meaning, ranked in one of the first three places by eighty-seven per cent of the students; next, but some way behind, came syntactic information (fifty-three per cent), closely followed by synonyms (fifty-two per cent), and then by spelling/pronunciation (twenty-five per cent), language variety (nineteen per cent), and etymology (five per cent). Béjoint comments that 'the interest in meaning suggests that for students the dictionary is basically an inventory of words with glosses', and that 'the dictionary is mainly used for decoding, since dictionary meanings are unlikely to be used for encoding activities' (p. 215).

Even though more than fifty per cent of the students claimed to consult their dictionary for syntactic information, it is not clear that they retrieve this information from the coding systems used. In answer to the question, 'Do you use the codes that indicate how a word should be used?', fifty-five per cent of the students said that they never used that information. This correlates with the response by eighty-nine per cent of the students that they had read the introductory matter to their dictionary only cursorily or not at all. It would appear that the compilers of these dictionaries have been presenting their target users with information that they do not use, or have not been able to access. And yet, only ten per cent of the students expressed themselves dissatisfied with their dictionaries, and seventy-seven per cent registered satisfaction. Perhaps they were not aware that the syntactic information was there, waiting to be accessed and made use of. Alternatively, it may be the case that users would be more likely to register dissatisfaction if information were excluded that they expected, rather than if information were included of which they felt no need. Béjoint notes, from the responses to another question, that 'the words that are looked up most often are those which typically cause difficulty when decoding' (p. 218). 'Conversely,' he writes, 'words which normally pose problems when encoding are seldom consulted' (ibid.).

Have the lexicographers got it wrong?

Two conclusions suggest themselves from the investigations of dictionary use discussed in this chapter. The first is that lexicographers consistently, or perhaps persistently, put into dictionaries certain kinds of information for which the vast majority of users have no need and would not miss if they were not included in dictionaries. Into this category would come grammatical information including part-of-speech labels, etymology, and perhaps pronunciation. If we were to investigate the range of words looked up in dictionaries we should probably find that we could dispense with a large number of the entries for 'common' words. This conclusion may have some validity as far as native-speaker dictionaries are concerned, though lexicographers might argue that, like the BBC, they have a duty to cater for the minority interests among their consumers as well as those of the majority. It is arguable, however, that these minority interests could be, or are already being, (better) catered for in the specialist dictionaries (e.g. of pronunciation, etymology), just as the BBC serves its minority interests with Radio Three.

In the case of the advanced foreign learners, however, this conclusion cannot be accepted, since it seems clear that the kind of syntactic and usage information so richly provided by the learners' dictionaries is essential if these dictionaries are to fulfil their undisputed function in aiding learners to encode acceptable sentences in the foreign language English. We must draw an alternative conclusion from the investigations, which relates to the users' perceptions of dictionaries and their capabilities of extracting information from them, i.e. what have been called users' 'reference skills'. Béjoint notes that for his students 'the dictionary is basically an inventory of words with glosses' (p. 215), a conclusion which would seem confirmed by the other surveys. In other words, there is no expectation on the part of dictionary users to find anything other than 'meanings' – and, of course, spellings – in a dictionary; consequently the other information is passed over. Moreover, the 'other information' is, to a greater or lesser extent, less accessible, since it is usually given either in abbreviation or in code. To access it, therefore, users have either to consult a table of abbreviations and/or codes, or to memorise the abbreviations/codes of their dictionary. Usually this means a careful reading of the often extensive front-matter of the dictionary (see Chapter 3, p. 36), and a conscious familiarisation through use of the abbreviations and/or coding system of the particular dictionary. There is, as Béjoint suggests, a case for teaching reference skills to dictionary users, and this case may well be extended to native-speaker users as well. It is to be hoped that this book has gone some way to perform this function for its readers.

Views of the dictionary

Let us now broaden the question of how 'the dictionary' is viewed by its users. We want to look at some general attitudes to the dictionary and compare these to how dictionary editors intend their publications to be viewed. One question that users often approach a dictionary with (see the Questionnaire at [1] and the discussion of the results) is: 'Does this word exist in English?' In playing word-games, or indeed in disputes about words that come up in ordinary conversation, recourse is had to a dictionary in order to determine whether a particular word exists in the language or not. 'The dictionary', as the repository of the vocabulary of a language, is the authoritative arbiter in such cases of uncertainty. There are two problems with viewing a dictionary in this way. The first and most obvious is that no dictionary, except the very large ones like the unabridged *Webster's Third New International Dictionary* with its four hundred and fifty thousand plus entries, comes anywhere near charting the whole of the vocabulary of English. A concise dictionary, which most people use, with its fifty thousand or so headwords, represents merely a selection of the words of English. Moreover, what is included will depend on the selection policy of the editors, e.g. the attention paid to scientific and technical terms, or to colloquialisms. The second problem with viewing the dictionary as the authority on what is and what is not a word in English arises with derived words, which are frequently the ones in dispute; e.g. is *weatherwise* (= 'from the point of view of the weather') or *jollify* (= 'make jolly') a legitimate word of English? A dictionary is frequently unable to provide an answer to this question, or it will provide the answer 'No', because such words are coined by applying highly productive derivational processes (see Chapter 2, p. 31), whose products cannot all hope to have an entry in dictionaries.

A second question that users consult a dictionary with is: 'Is this word a "proper" word of English?' This is a question about the words that you might use in a 'good style' of (usually) written English. Schoolchildren are sometimes told by their teachers to use only words that they find in 'the dictionary'; and this is usually an instruction to avoid colloquial or slang terms in their writing. The assumption here is that 'the dictionary' contains only those words that it is 'proper' to use in good written style, and that 'the dictionary' can therefore be appealed to as an authority on such matters. The problem with this view is that it rests on a complete misapprehension about the content of dictionaries. Lexicographers have long departed from the practice (if they ever wholly followed it) of including only those (uses of) words sanctioned by the writing of 'the best authors'. Consider the following quotation from the Preface to the first edition of the *Concise Oxford Dictionary* (ed. F. G. & H. W. Fowler, 1911):

> [2] . . . if we give fewer scientific and technical terms, we admit colloquial, facetious, slang and vulgar expressions with freedom, merely attaching a cautionary label; when a well-established usage of this kind is omitted, it is not because we consider it beneath the dignity of lexicography to record it, but because, not being recorded in the dictionaries from which our word-list is necessarily compiled, it has escaped our notice.

And that confession from an *Oxford* dictionary as well!

A third question, or set of questions, that users come to dictionaries with is seeking guidance on the way in which words should be used: 'How should I spell this word?', 'What is the right way in which to use this word?', and so on. Often such questions are wholly legitimate; a native speaker of the language is confused and bewildered, unsure about some point of usage, and consults a dictionary for help and guidance. And a dictionary will usually be able to provide the information required, e.g. that *occurrence* is spelt with two *c*'s and two *r*'s, that *inflatable* has no *e* in the middle, that *partake* is followed by *of* and is not synonymous with *participate in*, and that *affect* does not mean the same as *effect*. We are all linguistically insecure at certain points, and we need help in resolving our insecurity: that is what a dictionary is for. But it is going a step further to view 'the dictionary' as the ultimate arbiter and authority in all matters of linguistic usage, even where there is a genuine and widespread dispute or uncertainty. An honest dictionary should indicate that this is so. Look up, for example, the word *disinterested* in your dictionary. What does it have to say about the relation of this word to *uninterested*?

Most dictionaries would probably see it as part of their role to make a comment on disputed usages of this kind, but the comment may vary from an authoritative prescriptive pronouncement, such as in the *Heinemann English Dictionary* entry given at [3] below, to a note to the effect that the use of *disinterested* in the sense of 'uninterested' is deprecated by some speakers, as in the *LCED* entry given at [4] below.

> [3] Common Error. DISINTERESTED, UNINTER-ESTED both refer to a lack of interest, however *disinterested* describes impartiality or absence of selfishness, whereas *uninterested* suggests merely indifference or lack of sympathy.
>
> [4] **disinterested** *adj* 1 uninterested – disapproved of by some speakers 2 free from selfish motive or interest; impartial

Now consult your dictionary to see if it has anything to say about the alternative spellings of the following words:

> [5] ag(e)ing   enquire/inquire   judg(e)ment   medi(a)eval specialise/-ize

---

Most current dictionaries note these as alternative spellings and pass no comment on the preferability of one over the other. However, in four out of the five cases, the *Oxford Paperback Dictionary* makes implicit or explicit comment. It gives only the form *ageing* as the present participle of the verb *age*, the *Concise Oxford* (*COD*) includes both as alternative spellings. The *Oxford Paperback* comments on *enquire/enquiry* as follows: 'Although these words are often used in exactly the same way as *inquire* and *inquiry*, there is a tendency to use *en-* as a formal word for "ask" and *in-*

for an investigation.' Under *enquire, enquiry* the *COD* simply has the entry: 'see INQUIRE, INQUIRY'. The *Paperback Dictionary* has only the form *judgement*, whereas *COD* gives both spelling alternatives. The *Paperback* gives both forms of *medieval*, with the entry under *mediaeval* pointing the user to the alternative spelling; *COD* does the reverse. But the *Paperback Dictionary* lists only the form *encyclopaedia*, where the *COD* allows the alternative without *a* by giving the headword as 'encyclop(a)edia'. The *Paperback Dictionary* gives only the form *specialize*; again *COD* lists the alternative in -*ise*.

One final point about attitudes to dictionaries, before we turn to what dictionary compilers themselves think they are doing, brings us back to the first question on the Questionnaire at [1]. The answers to this question revealed that more than sixty-three per cent of the Birmingham Polytechnic students (over fifty-eight per cent of the English students, and seventy per cent of the Speech and Language Pathology students) owned dictionaries from the Oxford family (*Concise Oxford, Pocket Oxford*, and even the two volume *Shorter Oxford English Dictionary*). Indeed in many cases, students merely put down 'Oxford English Dictionary', not thereby meaning, I am sure, the great twelve-volume historical dictionary, but using that title to refer to their *Concise* or *Pocket* edition. It is a reflection of the prestige of the Oxford dictionaries in the public mind: the one great scholarly historical dictionary, which is actually quite useless as a current general-purpose dictionary, has imbued the name of 'Oxford' with a general aura of lexicographic authority. Quirk's students at University College London in 1972 had an even higher incidence of ownership of Oxford dictionaries: over seventy-five per cent (there were fewer alternatives available in 1972 than in 1986). It is not intended to denigrate the Oxford dictionaries, merely to point to them as the epitome of the exaggerated view that exists among the dictionary-buying public of the authority of 'the dictionary'.

How dictionaries view themselves

Consider the following extracts from the Prefaces and other front-matter of a selection of current dictionaries. What do they tell you about how the editors view the function of their dictionaries?

[6] *Longman Concise English Dictionary*: . . . the editors of this dictionary have attempted to provide . . . a reference work that gathers together over 70,000 current English words and expressions, from wherever in the world the language is spoken, gives a clear and concise account of their meanings, and offers guidance on the way in which they are used and pronounced . . . This is not a prescriptive dictionary; but it does set out to describe the prescriptions that exist in English.

[7] *Collins English Dictionary*: . . . the best and most up-to-date guide to the English language throughout the world . . . it records senses and uses of each word that are acceptable in the community as a whole . . . a fresh survey of the contemporary language . . . that sets out to reflect contemporary English as the international language it has become.

[8] *Pocket Oxford Dictionary*: In response to frequent requests from those who are concerned about standards of English that guidance be given on matters of disputed and controversial usage, the special markings introduced in the seventh edition of the 'Concise Oxford Dictionary' (**D** for disputed uses and **R** for racially offensive uses) are also adopted here in the smaller work . . . Two categories of deprecated usage are indicated by special markings: **D** (= disputed) indicates a use that, although widely found, is still the subject of adverse comment by informed users; **R** (= racially offensive) indicates a use that is regarded as offensive by members of a particular ethnic or religious group.

[9] *Chambers Twentieth Century Dictionary*: The . . . aim is to provide, in convenient and easily legible form, a comprehensive vocabulary aid for the present-day reader, speaker and writer of English . . . The new edition is catholic and broadminded in its noting of colloquialisms and slang terms, as

these may now be considered an integral part of spoken and written English.

[10] *Heinemann English Dictionary*: In so far as any concise dictionary can do so, we hope that this dictionary accurately reflects and illuminates the living, changing face of English in all its variety and richness . . . In cases where one headword is likely to be confused with another of similar spelling, function, usage, or pronunciation, a brief note labelled 'Common Error' distinguishes the two words.

You will note that in general the emphasis is on 'reflecting' and 'surveying', and that the editors intend their dictionaries to provide a 'guide' or 'aid' to the user, not a prescription, explicitly not in the case of the *Longman* editors. Dictionaries, then, apparently do not see themselves in quite the same way as their users often see them: the emphasis is on making a record of the language, reflecting its 'variety and richness', being 'catholic and broadminded' in their policy of inclusion, but at the same time providing 'guidance', presumably by appropriate labelling or usage notes, for the linguistically insecure. The extract from the front-matter of the *Pocket Oxford Dictionary* at [8] is of particular interest, however, as it provides an illuminating commentary on the tensions between the demands of the dictionary-buying public, or that part of it that requires authoritative statements from its dictionaries, and the lexicographer making an honest record of the language. It is noteworthy that this public turns to the Oxford dictionaries with its requests, and that the Oxford editors feel obliged to accede to the request, and so perpetuate the myth of the Oxford dictionary as the voice of authority in matters of linguistic usage in English. Incidentally, the 'uninterested' use of *disinterested* receives the '**D**' marking in the *Pocket Oxford*.

Dictionaries then do not claim to be the prescriptive authorities that many, if not most, of their users imagine them to be. The critics of the avowedly non-prescriptive *Webster's Third* in the 1960s, as we saw in Chapter 8, thought that the dictionary had failed to perform its public duty by not explicitly telling its users what constituted 'good' and 'proper' English. The voices of such critics can still be heard: consider the following extract from the Introduction to *Everyman's Good English Guide*, by Harry Fieldhouse (1984):

[11] It is fashionable to question whether there is even such a thing as correctness. Ordinary people undoubtedly believe there is. They regard dictionaries as authorities on what should be said and look them up in search of rulings. Modern dictionaries however have largely abdicated this role. Their compilers follow a 'descriptive' policy. Holding that language is a consensus, they are reluctant to prescribe one form rather than another. Their function, as they see it, is to record indiscriminately whatever is widely said, right or wrong. Scholarly as this may sound, it is rather like deducing the Highway Code from the way drivers and pedestrians actually behave. It turns dictionaries into guides to prevailing malpractices among the ill-informed.

We are back with the linguistic elitism of the advocates of an Academy in the eighteenth century, or perhaps with the authority of the 'best authors'. But the rules of language are not like those of road use, and dictionaries are not intended to be the 'highway code' of language. Bad driving causes accidents and costs lives; language is never 'bad' in this sense, though it may be inappropriate in a particular social context. The rules of language are social rules; they are more akin to whether you wear trousers with turnups or without, or whether you transport your peas to your mouth on the back of the fork or on the front. Some linguistic manners may be favoured by certain sections of society, but an offence against linguistic manners is not like contravening the Highway Code. And it is no task of the dictionary to prescribe linguistic social graces. Dictionaries attempt to record and describe the variety of linguistic manners that exist, rather than be manuals of linguistic etiquette.

Exercises

1. Keep a diary of your dictionary use over a period of a

month. Note when you look a word up, the occasion of the consultation, the word that you look up (common or 'hard'?), and the information that you expected to retrieve. Does your dictionary ever let you down because it does not contain either the item that you are looking up or the information that you require about an item? Is the failure a lack in dictionaries generally, or do you need to consult an alternative (perhaps larger) dictionary?

2. Does your dictionary have anything to say about the disputed or problematical usage of the following?

aggravate (= 'irritate' or 'make worse')   all right vs. alright   decimate   different from/to/than   due to vs. owing to   gaol vs. jail   media   principal vs. principle

3. Does your dictionary contain the following (derived) words? If any are included, is it as a run-on or as a separate entry?

anti-slavery   beautification   hacker   me-too-ism
openness   prettify   privatise   randomiser   re-employ
tankful

4. Discuss with your fellow students whether a dictionary ought ever to prescribe rather than describe, in advising, for example, against 'solecisms' (like irregardless, this phenomena), taboo words, sexist terms (like spokesman, poetess), racially or politically offensive words, etc.

CHAPTER 14

Not Alphabetical

Ask anybody for a definition of the word *dictionary* and the term *alphabetical* will most likely feature in it. The *Collins Pocket English Dictionary*, for example, defines it as 'a book of alphabetically listed words in a language, with definitions, pronunciations, etc.'. As we noted in Chapter 8, p. 112, however, dictionaries have not always been arranged alphabetically. Nor, it may be argued, is it necessarily the best arrangement for describing the vocabulary of a language. An alphabetical listing, after all, arranges lexemes in relation to each other in a purely arbitrary manner: adjacent entries in a dictionary rarely have any semantic relation, merely the accident of being in alphabetical series. The tradition of alphabetical arrangement probably developed because it served the convenience of reference: it is easier to find an item if it is located at the appropriate point in an alphabetical list, and nearly all reference books in daily use follow the dictionary tradition. Such an arrangement may be appropriate for telephone directories, but for dictionaries, as descriptions of the vocabulary of a language, it implies a view of vocabulary as a collection of unrelated words, whereas in fact the lexemes of a language, as we have seen, show many and various relationships.

In this chapter we shall consider some of the disadvantages of the alphabetical arrangement of vocabulary and investigate alternative ways of arranging and describing the lexemes of a language which take account of some of the semantic relations that we have discussed.

Disadvantages of alphabetical ordering

Although dictionaries are organised on the alphabetical principle, the extent to which the order, especially in respect of derived words, is strictly alphabetical, varies, as we noted in Chapter 3, p. 43, from dictionary to dictionary. Some dictionaries include all words derived by suffixation under the stem lexeme from which they are derived, so that, for example, *cranial* will be included under *cranium*, even though it precedes it alphabetically. Other dictionaries would place *cranial* before *cranium* in the appropriate position in the alphabetical order. Clearly we could put forward arguments for both ways of treating derived words: for ease of look-up a strict alphabetical arrangement is to be preferred; but the indication of derivational relationships is thereby sacrificed, and the user who wishes to know the appropriate adjective related to *cranium* will not find it easily in a dictionary with strict alphabetical ordering of lexemes.

This is even more the case where morphologically related items are not orthographically related. What, for example, are the adjectives corresponding to the nouns in [1]?

[1] church eye hand law sight skin son tooth
    two uncle

Apart perhaps from *law* and *uncle*, there is no observable orthographic relationship between the nouns in [1] and their corresponding adjectives, given in [2]:

[2] ecclesiastical ocular manual legal visual
    cutaneous filial dental double/dual avuncular

It is true that many of these adjectives belong to formal style and in some cases to the technical register. Nevertheless, if we regard the relationship of *church* to *ecclesiastical* as the same as that of *fanatic* to *fanatical*, and we consider this morphological relationship worth noting as part of the lexical description contained in a dictionary, then it is inconsistent to note the one because the words are orthographically related but not the other because there is no orthographical correspondence. The more consistent, though perhaps less revealing approach is to proceed strictly alphabetically without regard to morphological relatedness and indeed to ignore this altogether.

A far more serious disadvantage of an alphabetical arrangement for the lexical description of a language is the one hinted at right at the beginning of this chapter: lexical and semantic relationships between lexemes remain obscured by the arbitrary alphabetical principle of ordering. For example, consider the lexemes at [3]. What is the semantic relationship between them? If you are unsure, look the items up in a dictionary.

[3] braid   canvas   corduroy   denim   gabardine hessian   madras   muslin   organdie   percale seersucker   tweed   twill   velveteen   worsted

---

This orthographically disparate set of lexemes have in common that they all refer to different types of fabric. They differ in part in the kind of yarn used to manufacture them, as well as in the kind of weave and finish. If, for example, you wanted to know the meaning of *seersucker*, it would arguably be more enlightening to see its meaning displayed among those of all the other lexemes referring to fabrics, rather than by itself among the *S*-words of the dictionary. A technique of lexical analysis has been developed to provide just this kind of lexical description: it is called **lexical field analysis** or **semantic field/domain analysis**.

Lexical/semantic fields

The assumption underlying lexical field analysis is that lexemes can be grouped together into 'lexical fields' on the basis of shared meaning and that most if not all the vocabulary of a language can be accounted for in this way. The description of meaning, the definition of lexemes, is then undertaken within each lexical field and involves defining each lexeme in relation to the other lexemes in the field. Let us take as an example a small selection of the items in [3]:

[4] corduroy   gabardine   seersucker   tweed

These lexemes are defined in the *Longman Concise English Dictionary* (*LCED*) as follows:

[5] *corduroy*   'a durable usu cotton pile fabric with lengthways ribs or wales'

[6] *gabardine* 'a firm durable fabric (e.g. of wool or rayon) twilled with diagonal ribs on the right side'

[7] *seersucker* 'a light slightly puckered fabric of linen, cotton or rayon'

[8] *tweed* 'a rough woollen fabric made usu in twill weaves and used esp. for suits and coats'

The shared meaning of these lexemes is the fact that they all refer to fabrics: indeed 'fabric' would be the appropriate label to put on this lexical field. If you examine the definitions given in [5] to [8] you will notice that a number of semantic features recur in all the definitions: the kind of yarn used to make the fabric – cotton, wool, rayon, linen; the appearance or surface texture of the fabric – ribs, twilled, puckered; the thickness or weight of the fabric – durable, firm, light. These features, together with others no doubt *e.g.* the use of the fabric, as in the *tweed* definition), would be relevant to the definition of all lexemes in the lexical field of 'fabrics'. Definition then is by reference to a shared set of semantic features; lexemes in the field differ in the values that are assigned to these features. The semantic description of the fabric lexemes in [4] could be displayed by means of a matrix (compare Chapter 6):

[9]

	THICK-NESS	YARN	TEXTURE	USE
corduroy	durable	cotton	lengthways ribs	
gabardine	firm/ durable	wool/ rayon	diagonal ribs	
seersucker	light	linen/ cotton/ rayon	slightly puckered	
tweed		wool	rough/ twill weaves	suits coats

We should not imagine, however, that the vocabulary of a language is made up of a number of discrete lexical fields, in which each lexeme finds its appropriate place. Language can rarely be analysed into neat and logical compartments, least of all in the lexical area. First of all, there is no given set of lexical fields, no generally agreed set of labels. The number and composition of the lexical fields is the decision of the individual analyst. Many lexical fields may suggest themselves 'naturally', perhaps for example that of 'fabrics', but analysts may be faced with difficult decisions about the composition of a lexical field, especially where a lexeme would appear to fall in more than one field. The lexemes *wool* and *linen*, for instance, besides belonging to the field of 'fabrics', also belong to the field of 'yarns'. Do we therefore include 'yarns' and 'fabrics.' in a composite field? Or do we allow lexical fields to overlap, perhaps – so that we can account for lexemes that belong in more than one lexical field? These are questions that we do not want to go into any further here, but they serve to illustrate that what is true of linguistic descriptions generally is also true of lexical descriptions: they leak.

Let us turn our attention rather to further examples of lexical fields and how they contribute to the description of meaning. Consider the lexemes in [10]. Assuming that they constitute (part of) a lexical field, what label would you choose to designate the field?

[10] comet   galaxy   moon   nova   planet   satellite
     star   sun

You will no doubt agree that the lexemes in [10] have some shared meaning, that they could be said to be members of the same lexical field. They all refer to objects observable from earth in space, and we might use the term 'heavenly bodies' to label the lexical field. Now that we have determined the shared meaning, we need to look at the differences in meaning, defining each lexeme in terms of the others. You may find it useful at this point to look these words up in your dictionary and note their definitions.

There are a number of semantic features that suggest themselves in attempting to make a lexical description of the items in [10]. Firstly, we might make distinctions of generality: *galaxy* is more general in reference than *star*, and *star* is more general than *nova*. Secondly, we might distinguish those heavenly bodies which are naturally luminous (e.g. *star*) from those which are visible because they reflect the light of the first (e.g. *moon*). Thirdly, we might distinguish those heavenly bodies which move round others, from those which are themselves orbited.

Let us begin with the relationships of generality. Here we can invoke the semantic relation of **hyponymy**, which we mentioned in Chapter 6 (p. 91). Hyponymy refers to the semantic relation of inclusion: the meaning of a (more specific) lexeme is included in that of another (more general) lexeme. Thus *nova* is a hyponym of *star: star* is the **superordinate** term, and *nova* is the subordinate term in this semantic relation. Hyponymy is often important in displaying the semantic relations between items within a lexical field. Among the lexemes in [10] we can observe the following instances of hyponymy:

[11]

You will note that, although *galaxy* is more general in reference than *star*, it is not in a relation of hyponymy with it, since the meaning of *star* is not included in the meaning of *galaxy*: a star is not a kind of *galaxy*. Rather *galaxy* refers to a group or collection of stars: it is a **collective** rather than a **generic** term. We should also note that the lexeme *star* is used both in a generic sense to include *sun, nova* etc., but also in a specific sense, as a heavenly body distinct from sun, moon, etc. It should therefore appear twice in the hierarchy:

[12]

When we turn to the features of luminosity and orbiting, we find that there is a certain, though not complete correspondence with the sets established in [11]. The members of the *star*-group are naturally luminous, while those of the *satellite*-group are not, with the exception of *comet*. Similarly, the members of the *star*-group are non-orbiting but may themselves be orbited, while the members of the *satellite*-group are orbiting, though in the case of *planet* may also itself be orbited. We have not yet completely distinguished all the lexemes, e.g. *sun* from *star* (in its specific sense): a *sun* is a star that is the centre of a system, it has satellites orbiting it. We also need a feature to distinguish *nova* from *star* (specific): a nova is a star whose luminosity fluctuates, getting brighter and then fading. We might also note that we have ignored the meaning of *moon* referring specifically to the earth's satellite, and we are restricting *satellite* to its generic, natural (i.e. not man-made) sense.

It will now be possible to construct a matrix for the lexemes in [10]. In this case the features are binary, rather than multi-valued as they were for fabrics in [9].

[13]

	COLLECTIVE	GENERAL	LUMINOUS	FLUCTUATING LUMINOSITY	ORBITING	ORBITED
galaxy	+	−	+	−	−	−
star1	−	+	+	−	−	−
star2	−	−	+	−	−	−
sun	−	−	+	−	−	+
nova	−	−	+	+	−	−
satellite	−	+	+/−	−	+	+/−
planet	−	−	−	−	+	+
moon	−	−	−	−	+	−
comet	−	−	+	−	+	−

You will notice that two features (COLLECTIVE and FLUCTUATING LUMINOSITY) are necessary for only one lexeme each. Additionally, while this matrix distinguishes the meanings of these lexemes from each other, it does not tell us everything about the meanings of the lexemes, such as we might find in their ordinary dictionary definitions. For example, one striking fact about a *comet* is that its orbit round the sun is extremely elliptical. A matrix may not always be the most appropriate or the only appropriate way to present the description of a lexical field.

Let us investigate another lexical field to illustrate this point. Consider the verb lexemes in [14]. What lexical field would you say that they belong to?

[14] carve   chisel   chop   cut   hack   hew   prune
     segment   slice   trim   whittle

Here we are concerned with verbs of cutting, and 'cut' would be a suitable label for this lexical field. For each of the lexemes in [14], 'cut' is a component of their meaning; and this shared component justifies regarding them as (part of) a lexical field. What other semantic components or features are needed to distinguish the meanings of the lexemes in this field? Examine the lexemes in [14] and determine which features you think are necessary, looking them up in your dictionary, if you need to.

As far as the relation of hyponymy is concerned, the superordinate term is *cut* and all the other lexemes are hyponyms of this. They seem to fall into four groups, according to the semantic focus of the action of cutting. One group, containing just *chisel*, focuses on the instrument of cutting. A second group focuses on the nature of the action itself: *chop* implies a striking action and may almost be regarded as a superordinate term for the group; *hack* implies repeated rough blows with an instrument; *hew* implies regular, decisive blows; and *whittle* implies repeated small and gentle strikes that remove only small amounts of material at each strike. A third group focuses on the pieces that result from the cutting action; it includes *segment* and *slice*. And the fourth group focuses on the resulting appearance of the cutting: *carve* implies a sculpture or lettering, except in the 'slice' sense with meat: *prune* implies a resultant clean object and the removal of unwanted pieces; and *trim* implies a neat and tidy result.

It is difficult to envisage how we might display this analysis in the form of a matrix, either with binary components or with multi-valued ones. Besides, it hardly exhausts the semantic differentiation of the lexemes in [14]. A further important feature of the meaning of many of these verbs, for example, is the kind of syntactic object that they are usually accompanied by and the kind of material (wood, stone, etc.) that is implied by the verb action. *Hew* and *whittle*, for instance, are usually used of wood, *chisel* and *carve* of stone or wood; *prune* is usually used for trees and bushes, *trim* of hair and bushes, and so on. This collocational (see Chapter 7) information appears to be an important part of the meaning of lexemes in this field, but it would be difficult to capture with a component analysis. With lexical fields like this we can perhaps best display the semantic relationships by means of traditional dictionary definitions, with the lexemes grouped in such a way that their relationships are highlighted. This in itself is revealing and may indeed produce a more complete lexical description that would be appropriate for all lexical fields. Let us now turn to examine two attempts to produce dictionaries that arrange the vocabulary by lexical fields.

Non-alphabetical dictionaries

Probably the most famous attempt to group vocabulary by lexical fields is the *Roget's Thesaurus of English Words and Phrases* by Peter Mark Roget, first published by Longman in 1852, and appearing in many editions since. Roget had in fact begun to group words into fields in 1805, but it was only in retirement that he devoted himself fully to the compilation of what was to be published as the *Thesaurus*. A second edition followed in 1853 and a third in 1855, by which time the form of the *Thesaurus* had become fixed. Roget was still collecting additional words and phrases up to his death in 1869, when the editing task devolved to his son, John L. Roget, on whose death in 1908 it passed to the latter's son, Samuel R. Roget.

Peter Mark Roget wished in his *Thesaurus* to provide an alternative arrangement of words to the alphabetical ordering of dictionaries, 'according to the *ideas* which they express'. His aim was a very practical one. He wanted to provide a reference work to help people in composing written texts. One of the crucial difficulties in written composition he saw as being a lack of vocabulary. People often know what they want to express – they have the ideas – but they do not have the words, the range of vocabulary, by which to express them. In a sense the *Thesaurus* operates in the opposite way to a dictionary: a dictionary starts from words and tells you what ideas the words express, whereas the *Thesaurus* starts from ideas and tells you what words are used to express them. Additionally, Roget thought, the *Thesaurus* would enable people to develop their thoughts and ideas. On the argument that words are the means by which we think, he suggested that 'the review of a catalogue of words of analogous signification will often suggest by association other trains of thought, which, presenting the subject under new and varied aspects, will vastly expand the sphere of our mental vision' (from the Introduction).

While stressing his chief aim of 'practical utility', Roget devised a classification scheme of universal concepts. He envisaged that it would be applicable to all languages and talked of the usefulness of a 'Polyglot Lexicon' compiled according to the scheme. In Roget's scheme, there are six primary 'Classes of Categories', as follows:

[15] Abstract Relations   Space   Matter   Intellect
Volition   Affections

Each primary class is further divided and subdivided to produce a quite complex scheme of classification. For example, Class VI 'Affections' is first of all subdivided into:

[16] Affections Generally   Personal   Sympathetic
Moral   Religious

The 'Sympathetic Affections' category, for example, is then further subdivided into:

[17] Social   Diffusive   Special   Retrospective

At a final subdivision we now arrive at the sets of words themselves. The 'Retrospective Sympathetic Affections' category contains six sets of words, numbered 916 to 921:

[18] 916 Gratitude   917 Ingratitude   918 Forgiveness
919 Revenge   920 Jealousy   921 Envy

Each set of words in the *Thesaurus* is then organised by word-class. The entry for *921 Envy* reads as follows:

[19] **921. Envy. – N.** envy; enviousness &c. 'adj.';
rivalry; *jalousie de métier.*
V. envy, covet, lust after, crave, burst with envy,
regard with envious eyes.
Adj. envious, invidious, covetous; *alieni appetens.*

You will note that, beyond the word-class division, there
is a further implicit subdivision by means of the punctua-
tion. Closely related subsets of words are bounded by
semi-colons.

In the body of the *Thesaurus* the entries are arranged
on the page partly in double columns, partly in a single
column. Words are grouped together into sets on the basis
of 'analogous signification' or loose synonymy (see Chapter
5, p. 67). Where it is appropriate, sets that are considered
opposite in meaning are placed side by side in two columns
on a single page. For example, *916 Gratitude* has next to it
917 Ingratitude; *918 Forgiveness* is placed by *919 Revenge*; but
920 Jealousy and *921 Envy* are in the single-column format,
because they have no antonymous sets. We can see then that
the semantic relations of synonymy and antonymy play an
important role in the organisation of *Roget's Thesaurus.*
While there is a hierarchical scheme of categorisation, actual
lexemes are provided only for the lowest categories in the
scheme, so that we cannot really speak of a relation of
hyponymy operating, since there is no explicit ordering of
specific terms in relation to more general ones. However,
within sets we do sometimes find that more specific items
are grouped separately from their more general counter-
parts; for example, in the *366 Animal* set, *beast, brute* and
creature come together in one group, while *horse, cow, sheep,*
etc. are grouped together elsewhere in the entry. But there
is no systematic attention to the relation of hyponymy.
Indeed, Roget says that he has deliberately omitted 'a
multitude of words of a specific character' because they have
'no relation to general ideas' and do not therefore come
within the scope of the *Thesaurus* as he conceives it. You
will note, though, from the entry for *921 Envy* at [19] that
he includes not only single words, but phrases and
expressions as well, i.e. idioms and firm collocations.

Roget anticipated that the *Thesaurus* would be used in
two ways. On the one hand he expected people to come
with a general idea of what they wanted to write about and,

beginning with the table of categories, would find the appropriate section of the *Thesaurus* which would provide them with the stores of vocabulary items from which they could choose those suited to their purpose. On the other hand he recognised that people might also come with a word that they wished to set in the context of similar words. And for such likely users he provided a copious index to the *Thesaurus*. Indeed the Index takes up almost half of the book. Where a word has several senses and is therefore placed in more than one set, each sense is indicated in the Index with reference to the relevant set, e.g. for *lukewarm*:

[20] **lukewarm**

temperate 382	(under *Heat* in Class III Matter)
irresolute 605	under *Irresolution* in Class V Volition
torpid 823	(under *Insensibility* in Class VI Affections)
indifferent 866	(under *Indifference* in Class VI Affections)

Roget's Thesaurus is then what has proved to be an enduring attempt to organise the vocabulary of English into lexical fields, before that term was ever articulated; but all we are given is simply a set of words, with no definitions and no attempt to differentiate their meanings one from another.

A second, more recent attempt to describe the vocabulary of English in terms of lexical fields is the *Longman Lexicon of Contemporary English* by Tom McArthur (1981), a companion volume to the alphabetically arranged *Longman Dictionary of Contemporary English* (*LDOCE*), and like this aimed primarily at the advanced foreign learner of English. It differs from Roget's *Thesaurus* in a number of significant ways. However, in basic layout it is similar: the main body of the dictionary, containing the lexical fields, is followed by an alphabetical index of the approximately fifteen thousand entries, where the pronunciation is given.

The most striking and significant difference between McArthur's *Lexicon* and Roget's *Thesaurus* is the basis on which the lexical fields are determined. The philosophical scheme of 'universal concepts' is absent from the *Lexicon*. Pragmatic considerations have determined the selection of the fourteen broad fields of the *Lexicon*: their usefulness to the intended users of the *Lexicon*, and consequently their relevance to the everyday life of the modern world. The fields are listed 'A' to 'N' as follows:

[21] A Life and Living Things
B The Body: Its Functions and Welfare
C People and the Family
D Buildings, Houses, the Home, Clothes, Belongings, and Personal Care
E Food, Drink, and Farming
F Feelings, Emotions, Attitudes, and Sensations
G Thought and Communication, Language and Grammar
H Substances, Materials, Objects, and Equipment
I Arts and Crafts, Science and Technology, Industry and Education
J Numbers, Measurement, Money, and Commerce
Entertainment, Sports, and Games
L Space and Time
M Movement, Location, Travel, and Transport
N General and Abstract Terms.

Each of the broad fields is then subdivided into 'sets'. In field D, for example, there are eight sets, as follows:

[22] Architecture and Kinds of Houses and Buildings
Parts of Houses
Areas Around and Near Houses
Residence
Belonging and Owning, Getting and Giving
Furniture and Household Fittings
Clothes and Personal Belongings
Cleaning and Personal Care

Each of these sets is then further subdivided into groups of related lexemes. For example, the set 'Residence' contains nine subsets (D60 – D68) as follows:

[23] D60 verbs: living and lodging
D61 verbs: accommodating people
D62 verbs: camping and settling
D63 nouns: persons residing, lodging and settling
D64 nouns: persons owning and occupying houses
D65 nouns, etc.: home

D66 nouns: accommodation and board
D67 nouns, etc.: shelter
D68 nouns: camp

The subset D63, for example, contains the following lexemes:

[24] resident   dweller   occupant   occupier   lodger
     boarder   settler   pioneer   squatter   inhabitant

However, and in this respect also McArthur's *Lexicon* differs significantly from Roget's *Thesaurus*, each of these subsets is not a mere list of words. Each word has a full dictionary-type entry, with definitions, examples, grammatical information, and indications of stylistic and register constraints. Indeed, the same grammatical coding is used as in *LDOCE* (1978). It is thus possible, for any lexical field or subfield, to compare in detail the meanings of all the lexemes. This produces not only a richer lexical description, in view of the fact that the meaning of a lexeme is determined by the meanings of semantically related lexemes, but also a more useful tool for the foreign learner, who often has difficulty in understanding the semantic boundaries between lexemes in another language. It is to be hoped that McArthur's *Lexicon*, of relatively modest coverage, will prove to be a trail-blazer for more lexical-field dictionaries.

One interesting point, that may count against the lexical-field dictionary, needs to be noted. Where a lexeme has different senses that belong in different lexical fields it is not possible to gain easily a view of the overall meaning of the lexeme. For example, the verb *accompany* is entered in two separate places in McArthur's *Lexicon*: in K32, in its musical sense, along with *play, sing, whistle, dance, perform* and *tune*; and in M53, in its general sense, along with *take, lead, guide, conduct* and *escort*. A verb like *take* is entered in nine different subfields. The index does show that such lexemes have multiple entries and gives a general indication of what each sense is; but there is no one place where the lexeme receives a composite description. Perhaps we should regard lexical-field dictionaries as complements to alphabetical dictionaries, rather than as replacements of them. Alternatively, we may wish to revise our whole approach to lexical description and be ready to recognise far more instances of homonymy and fewer of polysemy. That is to say, we would regard *accompany* in the musical sense and *accompany* in the general ('escort') sense as different (homonymous) lexemes, not as different senses of the same lexeme. Perhaps it is the historical bias of the alphabetical dictionary, with its reliance on etymology to distinguish homonyms from polysemes, that conditions us to view this as a case of polysemy. A lexical-field dictionary would provide an alternative lexicological perspective.

Thematic lexicography

What we have been discussing in this chapter is arguably a different kind of lexicography from that which has been traditional at least since the early eighteenth century. The only available examples of this alternative, 'thematic', lexicography are Roget's *Thesaurus* and McArthur's *Lexicon*, and it is only the latter which provides anything like a full lexicographical description of the items it includes.

This alternative lexicography arises out of new insights and approaches in lexicology (see Chapter 16, p. 243). In lexicology the emphasis is no longer on the consideration of words as isolated lexemes in the vocabulary of the language, to be treated one by one in terms of their forms and meanings. Much rather the emphasis is on the ways in which the vocabulary hangs together as a system or as a system of systems, with each lexeme having formal and more especially semantic links with many other lexemes in the vocabulary. Indeed it is considered impossible to make an adequate lexical description without invoking these links and relationships.

Lexicography lags behind in the application of these insights and approaches. It is a tradition with considerable in-built inertia and conservatism, imposed by public and publishers alike. But if future dictionaries are to reflect the advances made in semantic and lexical analysis and description, and so be more adequate and useful as descriptions of vocabulary, then more attention will have to be paid to constructing them on a thematic rather than on an alphabetical model.

Exercises

1. Look through the section of your dictionary with words beginning with **ga-**. List those concerned with BUILD-INGS. Make a preliminary analysis for a lexical-field description.

2. What features would you consider it necessary to identify in order to distinguish the meanings of the following lexemes in the lexical field of 'sending':

   send   transport   convey   deliver   dispatch

   It may help you to look up their definitions in a dictionary.

3. Consider the following set of lexemes from an entry in Roget's *Thesaurus*. What semantic feature do they have in common? And in which section would you expect to find them in the *Thesaurus*?

   *reminiscence   recognition   recurrence   recollection
   rememoration   retrospect   after-thought*

4. Consider the following set of lexemes from Section E on Food, etc. in McArthur's *Lexicon*. What do they have in common semantically? And how do they differ in meaning?

   cut   joint   slice   cutlet   steak   chop   fillet   rasher

CHAPTER 15

The Craft of Lexicography

Dictionary compilation has come a long way since the time when Samuel Johnson laboured with his six clerks in his chamber in Gough Square for seven years to produce the *Dictionary of the English Language* in 1755. It has come a long way indeed since James A. H. Murray in the 1880s worked alone in the 'Scriptorium' in his garden in Mill Hill, sorting with the help of his children the mass of citation slips sent in by the world-wide army of voluntary readers, and writing the articles that were to form part of the *Oxford English Dictionary*. There are still today individuals who undertake virtually single-handedly, or perhaps in pairs, the compilation of a dictionary, such as Tom McArthur and the *Longman Lexicon of Contemporary English* (though most of the material was derived from *LDOCE*), or the pairs responsible for the *Heinemann English Dictionary* (K. Harber and G. Payton) and the *Oxford Children's Dictionary* (John Weston and Alan Spooner). But these are quite modest undertakings, and most dictionaries of any size involve in their compilation an editorial team, backed up by a number of specialist consultants to advise on technical registers, national varieties and the like, as well as probably these days a computer, which would be used for a whole variety of tasks, from data storage and retrieval to checking and type-setting procedures.

Samuel Johnson self-deprecatingly defined *lexicographer* as 'a harmless drudge', but this greatly underestimates the skills, acumen and judgement required in a dictionary enter-prise. Patrick Hanks, himself an experienced lexicographer, writing the introductory article 'Meaning and Grammar' in the *Collins English Dictionary* (first edition, regrettably omitted from the second), portrays the (ideal) lexicographer rather as a polymath:

[1] . . . the lexicographer must be widely read; he must have a deep understanding of the terms he is defining, based on an extensive reading of the literature in which those terms occur. A vast citation file is no substitute for the judgement of the lexicographer. (p. xxxv)

Not only does the lexicographer need this breadth of knowledge and experience in order to compose succinct and lucid definitions, but it is also needed for the many general decisions that need to be taken in the course of compiling a dictionary. Some of these decisions will affect the layout of the dictionary: the number and size of the type-faces to be used, whether the swung dash will be used to save repeating a headword within a dictionary entry, whether strict alphabetisation will be imposed or all derived words given as run-ons, and so on. Such decisions will have a major effect on the one hand on how accessible a potential user will find the information in the dictionary, and on the other on how much information will be able to be contained within the limited compass of the projected work, and on the price at which it will be sold.

The content of the dictionary will be affected by other general decisions of the lexicographer: selection of items to be included in the dictionary, the information to be provided for each headword, the transcription system used for representing pronunciation, whether to include abbreviations or foreign words and phrases in the body of the dictionary or in separate appendixes, and so on. Such decisions concerning content and layout will not, for the most part, be the free and unfettered choice of the lexicographer. The overall purpose of the dictionary will constrain content and layout: whether it aims to be a record of the current language, as comprehensive as possible within the proposed size; or whether it aims to be a reference handbook for a specific set of users (schoolchildren, teenagers, foreign learners). The anticipated size, both page format and number of pages, will constrain at least the layout of the dictionary. A further constraint, especially on content, will be the general public expectation of what a dictionary should contain and how far a publisher is prepared to introduce innovations in dictionary design that go beyond that expectation.

We have touched on some of these points in the preceding chapters of this book. Let us now consider some of the tasks that lexicographers have to undertake in **compiling** dictionaries.

Amassing the data

It would be conceivable for a team of lexicographers to get together and compile a dictionary from their combined knowledge of and intuitions about the vocabulary of the language, without reference to any other source of information. The result might well be a very limited dictionary with a lot of gaps. The knowledge and intuition of lexicographers does play a part in the compilation of dictionaries, however, though no editorial team would acknowledge it as one of their primary sources of data and information. It is difficult, if not impossible, for a describer of a language to suppress all private judgements about the language, especially if the describer is a native speaker of the language.

Besides the generally unacknowledged intuitions, lexicographers have two main sources of data for their task of compiling dictionaries. One is previously published dictionaries, and the other is original source-text material. Most lexicographers probably use a combination of both data-sources. Samuel Johnson worked from an interleaved copy of the 1736 edition of Nathaniel Bailey's *Dictionarium Britannicum*, but supplemented this with excerpts from his own extensive reading. The *Oxford English Dictionary* (*OED*) on the other hand relied almost exclusively on excerption: the provision by the voluntary readers of slips of paper containing sentences from what they had read, considered to illustrate a sense of a particular word in such a way as to be almost self-defining. These citations from source-text material provided the raw data for the description of a lexeme, both of its various senses and of the historical development of its form and meaning.

When F. G. and H. W. Fowler, inexperienced in lexicography, embarked on the compilation of the *Concise Oxford Dictionary* (first published in 1911), they resolved to follow the same principles as were being used for the *OED*, except that the source of their citations now became the *OED* itself, at least for the letters 'A' to 'R', which was all that had been published at the time, as they noted in their Preface:

[2] This procedure – first the collection of sentences from all possible sources as raw material, and then the independent classification – we have often followed even in that part of our book (A–R) in which the O.E.D., with senses already classified and definitions provided, was before us, treating its articles rather as quarries to be drawn upon than as structures to be reproduced in little . . .

This would seem to imply that in part they did draw on the *OED* for sense divisions and definitions as well. For the remaining part of the dictionary (S–Z) they 'collected the quotations given in the best modern dictionaries' and then 'added to these what we could get either from other external sources or from our own heads'. After that, they claimed that they 'then framed our articles, often without reference to the arrangement that we found in any of our authorities', and presumably sometimes with reference to these authorities. Overall, then, while they obtained their data from secondary sources (other dictionaries) it was citation data rather than whole entries, though their reliance on previously published dictionaries conditioned the range of items that they included (see the quotation at [2] in Chapter Thirteen).

Philip Gove in the editorial Preface to *Webster's Third New International Dictionary* points both to existing dictionaries and to a wide-ranging policy of excerption as the source of the data underlying that dictionary:

[3] . . . the definitions in this edition are based chiefly on examples of usage collected since publication of the preceding edition [i.e. 1934]. Members of the editorial staff began in 1936 a systematic reading of books, magazines, newspapers, pamphlets, catalogs, and learned journals. By the time of going to press the collection contained just under 4,500,000 such new examples of recorded usage to be added to the more than 1,665,000 citations already in the files for previous editions. Further, the citations in the indispensable many-volume *Oxford English Dictionary*, the new citations in Sir William

Craigie's four-volume *Dictionary of American English* and Mitford M. Mathews' two-volume *Dictionary of Americanisms*, neither of which was available to the editors of the preceding edition, and the uncounted citations in dozens of concordances to the Bible and to the works of English and American writers and in numerous books of quotations push the citation background for the definitions in this dictionary to over ten million . . .

Again, the data that is collected comprises citations, from which the entries in the dictionary are then composed, involving decisions on division into senses and the content of definitions (see further below).

For both *OED* and *Webster's Third* the source data is a collection of citations. In a sense this data is no longer 'raw': it has been excerpted from texts. The extent to which either collection of citations is representative of the language as a whole depends on the range of texts from which they have been drawn and the insights and judgements of the excerpters. The texts underlying the *OED* citation collection were predominantly literary, though Murray also included citations from his newspaper reading, and it is possible to discern when he changed his daily paper. The texts underlying the *Webster's Third* citation collection represent a much wider range of language varieties, as the quotation from the Preface at [3] indicates. However, for neither citation collection can we be sure that the excerption has been totally comprehensive and systematic, since it relied, in the case of *OED* on largely amateur though in the case of *Webster's Third* presumably professional, human 'lexicologists', who more than likely missed many citations that might have changed the way in which entries in the dictionary were composed.

A possible solution to this dilemma would be to construct a representative corpus of texts, which would in itself be difficult enough and would presumably amount to several million words. An index would then need to be made of every lexeme in the corpus. A selection of the lexemes would be made for the size of dictionary intended. And a concordance would be constructed for all the selected lexemes. In this way each occurrence of all the lexemes selected would be included as a 'citation', and the dictionary

229 Selection and presentation 229

entries would be composed from a comprehensive citation collection from the representative corpus. Such an undertaking – indexing and concordancing – involving much of Johnson's drudgery, would take many editor-years of work. With modern computers, however, the feasibility of such an approach to data collection begins to become a reality.

Selection and presentation

Except in the case of dictionaries that aim to be truly comprehensive, like *Webster's Third*, a decision has to be taken at some point on which lexemes to include and which to omit. The decision may be taken either before the collection of the primary data, so that only evidence for the selected headwords is collected; or it may be taken after the collection of citations, when judgements can be made of frequency of occurrence and typical contexts of use. For derivative dictionaries (i.e. concise and pocket versions of larger works) the decision may be made quite early on, though even here items of more recent coinage than the date of publication of the larger dictionaries may be included in the smaller ones.

Lexicographers must therefore develop a selection or inclusion policy for each dictionary. The *OED* had a policy of concentrating on what were called the 'common words' at the core of the vocabulary; more peripheral vocabulary items were regarded as those that belonged to specialist technical registers or to regionally restricted varieties of the language (see the General Explanations to the *OED*). A similar policy seems to have been followed by the Fowlers in the first edition of the *Concise Oxford Dictionary* (see the quotation at [2] in Chapter 13). Modern dictionaries, on the other hand, seem to be concerned to have an inclusion policy that takes full account of the technical registers, since it is frequently lexemes from these areas that dictionary users need to look up, because science and technology impinge more on the everyday life of today than they did seventy or a hundred years ago. Perhaps the boundaries of the 'common core' have extended further and further into the peripheral areas. The inclusion policy of the *Collins*

English Dictionary (*CED*) is expressed in the Publisher's Foreword as:

[4] We thought it was essential that the dictionary should cover all the spoken and written English that is likely to be required by any but the most highly specialized users . . . We have . . . included . . . an exceptionally wide range of scientific and technical terms, in order to provide a truly comprehensive and useful dictionary . . .

Equally, dictionaries' policy on regionally restricted vocabulary is wider than it used to be. While many dictionaries still seem to include very few dialect words (*CED* is a notable exception), most would claim to recognise the 'international' status of English and include items from other national varieties. These, after all, have a greater currency, e.g. in the literature and films of various parts of the English-speaking world. Indeed, the Oxford family has recently produced a New Zealand version of the *Pocket Oxford Dictionary*, based on the original *POD* with the inclusion of specifically New Zealand vocabulary. Whether this volume is a one-off, the offspring of the New-Zealand-born current editor of *OED*, R. W. Burchfield, or whether it is the precursor of further regionally based dictionaries remains to be seen.

Besides selecting the headwords to be included and amassing the data on which the entries will be based, the lexicographer must decide what information to give for each headword and how it is to be presented. Unless the dictionary is aimed at a special group of users, such as children or foreign learners, much of the information to be given for each headword is dictated by the tradition of lexicographic practice: above all definitions (see below), but also pronunciation, part-of-speech and etymology. A greater divergence of practice is discernible, however, when it comes to style and register information. *Webster's Third*, for example, deliberately eschews register labels; in the words of the Preface, 'it depends upon the definition for incorporating necessary subject orientation'. Many other dictionaries find it more economical or more satisfactory to mark lexemes or senses of lexemes with an appropriate label.

The question of presentation, not just in terms of layout and the use of different type-faces, is very important, since it affects the accessibility of the information in the dictionary. This question includes the choice of transcription system for representing pronunciation (International Phonetic Alphabet or a home-grown one), and the extent to which abbreviations are used. Lexicographers are under pressure from publishers to abbreviate as much as possible, in order to keep down the size and cost of a dictionary, or alternatively in order to be able to put more information in. The Fowlers in the first edition of *COD* aimed to save space by:

[5] the curtest possible treatment of all [words] that are either uncommon or fitter for the encyclopaedia than the dictionary, and by the severest economy of expression – amounting to the adoption of telegraphese – that readers can be expected to put up with.

There are conventional abbreviations for the parts-of-speech ('n.', 'v.', 'prep.', etc.) that every user is expected to understand. In the etymological information, users are often expected to be familiar with abbreviations like *OE*, *ON* and *MHG* (for 'Old English', 'Old Norse', 'Middle High German', respectively); though it is interesting to note in this regard that in the Collins family, while *CED* does not use such abbreviations in its etymologies, the smaller dictionaries do. With other labels, such as style and register labels, dictionaries either abbreviate in such a way that the full form is easily suggested (*fml* for 'formal', *colloq.* for 'colloquial', *Med.*, for 'Medical', *Biol.* for 'Biology') and/or they provide a list of abbreviations in the front matter, or they do not abbreviate. *CED* is generally very sparing in its use of abbreviations; among register labels, for example, '*Chem*(istry)', '*Med*(icine)' and '*Path*(ology)' are abbreviated, while '*Anatomy*', '*Logic*' and '*Military*' are not.

Other issues of presentation would include those we have treated elsewhere in this book, e.g. how to treat derivatives – as run-ons or separate entries or both, depending on the meaning relation with the headword – how to treat idioms, phrasal and prepositional verbs, and other fixed expressions or multi-word lexemes, and how to treat irregular inflections. The question of how these various kinds of information are ordered in an entry is also important. But let us now turn to the central kind of dictionary information: meaning.

Dealing with meaning

Before writing the definition of a lexeme decisions need to be taken on the number of senses that a lexeme has and in which order to present the senses. Such decisions will also involve the criteria to be applied in drawing the line between polysemy and homonymy: how unrelated does a 'meaning' have to become before it is regarded as belonging to a separate lexeme with the same form? Will etymology be invoked as the sole and decisive criterion, or will the lexicographer recognise homonyms on semantic grounds and thus be required to make sensitive semantic judgements? And how will lexemes with multiple class membership be treated, as homonyms or as different senses of one lexeme? Such questions may be answered in pragmatic rather than linguistic terms: if the editors are pursuing a policy of relieving the density of entries and having more, shorter entries rather than fewer, longer ones, then there is likely to be a larger number of homonyms as well as the tendency to treat derivatives (if they need to be defined) and idioms as separate headwords. The appearance of the dictionary page and the supposed accessibility of the information may be deciding factors, rather than etymological or semantic considerations.

When it comes to the ordering of senses, lexicographers differ considerably in their practice. Some apply an historical ordering to the senses, as does *Webster's Third*, for example:

[6] In definitions of words of many meanings the earliest ascertainable meaning is given first. Meanings of later derivations are arranged in the order shown to be most probable by dated evidence and semantic development. This arrangement applies alike to all meanings whether standard, technical, scientific, historical or obsolete. (Preface)

The Fowlers, on the other hand, in the *Concise Oxford*, even though they used the *OED* as a source, did not consider an historical ordering of senses generally to be appropriate for their dictionary:

> [7] Occasionally, when a rare but still current sense throws light on the commoner senses that follow or forms the connecting link with the etymology, it has been placed at the beginning; but more commonly the order adopted has been that of logical connexion or of comparative familiarity or importance. (Preface)

A similar, though more differentiated policy is followed in *CED*: senses are grouped by part-of-speech; the first sense is usually 'the one most common in current usage', unless one of the other senses represents a 'core meaning' that illuminates the meaning of the other senses. Within the block of senses for a part-of-speech:

> [8] . . . closely related senses are grouped together, technical senses generally follow general senses; archaic and obsolete senses follow technical senses; idioms and fixed phrases are generally placed last. (Guide to the Use of the Dictionary)

We might consider the real craft of lexicography to reside in the writing of the definitions: the choice and arrangement of words in order to characterise the meaning of other words. But here again particular lexicographic traditions and policies will determine how the definitions are constructed (compare Chapter 9, p. 131): whether to attempt complete analytical definitions (*Webster's Third*), whether to emphasise substitutability of definition for head-word (*CED*), whether to have 'scientific' or encyclopaedic definitions of plants, animals, etc. (*CED*), and so on. Even within these constraints, however, it is arguably here that the skill and experience of the lexicographer is paramount, in crafting concise, succinct and lucid definitions – a craft to be learned, trained and developed (see R. Ilson (ed.), 1986).

Complementary to the definitions are the illustrative examples. It is equally a skilled decision to recognise when an example would be helpful in elucidating a definition, as well as to compose an appropriate example or select an apposite citation. The Fowlers, for instance, laid great store by the inclusion of illustrative sentences (mostly from literature) in the *Concise Oxford*. In the Preface to the first edition they mention as a peculiarity of the dictionary:

> [9] . . . the use, copious for so small a dictionary, of illustrative sentences as a necessary supplement to definition when a word has different senses between which the distinction is fine, or when a definition is obscure or unconvincing until exemplified . . .

The other, particularly important function of illustrative examples is to show a typical collocation or a typical syntactic structure into which (the sense of) a lexeme enters.

It would be appropriate at this point to draw attention to a study by Professor John Sinclair, which calls into question much of the traditional lexicographic practice that we have been discussing. It is reported in an article entitled 'Lexicographic Evidence', which appeared in *Dictionaries, Lexicography and Language Learning*, edited by Robert Ilson (1985). Sinclair compares the definition of the lexeme *decline* in *CED* with the evidence collected by concordancing a computer corpus of written and spoken English of more than seven million words (the 'Birmingham Corpus'). He first of all notices that a number of the derivations of *decline*, some entered as separate headwords in *CED* (e.g. *declension, declinate*) and some as run-ons (e.g. *declinable, decliner*), do not occur at all in the corpus; and he comments (p. 86):

> [10] So a word which does not occur at all in over 7 m. words of general current English does not have a strong claim to be in any dictionary of it.

Similarly, some of the senses given for *decline* in *CED* are not found in the corpus, and for two of the occurring senses, Sinclair questions whether the dictionary has not overdifferentiated the senses.

Perhaps the most disturbing conclusion of Sinclair's study, though, relates to the way in which dictionaries define lexemes as wholes, without regard to the relation of word-form and meaning. The *CED* entry divides the meaning of *decline* into five verb senses and four noun senses. Sinclair suggests that, besides the 'refuse' sense of the verb, there are two senses shared by both verb and noun: (1) grow smaller, diminish; movement downward, diminution; (2) deteriorate gradually; gradual deterioration or loss. The two senses are not, however, always distinguishable, and a number of indeterminate examples were found in the corpus. Sinclair notes further that the 'refuse' sense is entirely verbal and associated with the past tense form *declined*. For the other pair of senses, the form *decline* is usually a noun and tends to be associated with the 'deteriorate' meaning; *declined* is verbal and associated mainly with the 'diminish' meaning, as is *declining* – the other commonly occurring form, but whose function is predominantly adjectival. Sinclair suggests that such study of 'instances' will bring greater precision to lexicography and opens the way for considerable development of the craft.

This is an approach to lexicographic description that needs to be taken seriously. Such corpus-based approaches to linguistic description have not been possible until recently; but computer technology has now developed so that large corpuses of data can be stored and searched easily. We may query with Sinclair whether his seven-million-word corpus really is large enough and representative enough; but we must acknowledge that his approach of taking all the separate instances of all the forms of a lexeme is radical and innovative, and it brings a new perspective on lexicographic description. The traditional approach is to treat lexemes without regard to their different grammatical forms, and to recognise different senses of a lexeme on the basis of a limited number of citations, often from highly literary contexts. Sinclair's proposal is that we collect all the instances of the different forms of a lexeme from a repre-sentative corpus of texts (written and spoken), and that we make generalisations about the meanings of the lexeme from the instances collected, taking into account that different meanings may tend to be associated more strongly with particular forms. Such a consistently empirical approach applied to the whole range of entries in a dictionary would provide us with a more accurate record of how lexemes are used in the contemporary language generally.

Using a computer

Sinclair suggests one way in which a computer might be used to improve dictionary making: reading large text corpuses for the citations on which dictionary entries are built. But computers have been used extensively in lexicography already, if for nothing else, then at least to relieve some of the drudgery in compiling and sorting the collection of citations, and perhaps in typesetting as well. The potential for the application of computer technology to lexicography, as Tom McArthur discusses it in his book *Worlds of Reference* (1986), has probably nowhere near yet been fully realised.

The dictionary publisher's dream might be the application of computers not only for the excerption of citations but also for the construction of the entries themselves, in which a computer program would establish the senses and compose the definitions. But these are tasks that we have associated particularly with the 'craft' of lexicography, and the prospect of programming a computer to undertake them seems remote if not impossible at the present. The computer is being used, however, as in many other areas, for the fast and efficient storage and retrieval of information, both the basic data – word-lists, citations, etymologies, etc. – and the constructed entries. Indeed, it is possible for several lexicographers (pronunciation editors, definers, etymologists) to work on an entry simultaneously and be aware of what each other is doing. It also means that the lexemes or senses of lexemes can be coded for formality level, lexical field, syntactic class and the like, thus enabling items to be retrieved according to a variety of characteristics. It will make the compiling of thematic dictionaries (see Chapter 14) more feasible.

A further important purpose for which computers are being used in dictionary making is checking; for example, checking that all the words used in the definitions are themselves defined in the dictionary. The *Longman Dictionary of Contemporary English*, which used a restricted two-thousand word defining vocabulary (see Chapter 12, p. 187), was checked by computer to ensure that the definitions contained only items included in the defining vocabulary.

The computer has also been used to check that adequate coverage has been given to specialist terms from technical registers. Such checking procedures, which a computer is able to perform speedily and efficiently, ensure that dictionaries are more accurate and consistent than previously.

There are considerable advantages to having the completed dictionary stored in computer form. It ensures that the time between compilation and publication is reduced, since it means that it can be proofread substantially via visual display units while still in the machine and while last-minute revisions are being incorporated, and that through computer typesetting page-proofs can be available more quickly for checking the use of typefaces and page layout. It also means that subsequent editions do not become major undertakings in themselves and that constant updating should enable new editions to be produced more frequently. Indeed, a substantial recasting of the dictionary, e.g. in a thematic format, will become manageable in a computerised form.

Dictionaries in the form of books will no doubt be with us for a long time to come. But the era of the electronic dictionary cannot be far away, though the advantage will probably be for scholars rather than the general public. The *OED* is being computerised; *CED* exists in machine-readable form. If such dictionaries, and more, are included in data-bases that can be accessed through local computers via the telecommunications network, then scholars will have instant access to a greater and more up-to-date range of lexicographical information than at present. And it may not be too long before the general public will be able to purchase such data-bases in the form of compact discs and access them with home computers through a compact disc player. The on-line dictionaries will be continually updated; the home user will have to exchange the compact disc every so often for the latest edition. Looking further into the future, perhaps, we might envisage an interactive dictionary, which will be capable of being updated by its users, who will be able to register new lexemes or new uses of existing lexemes to the on-line dictionary through their own computer terminals.

Exercises

1. Examine your dictionary (or dictionaries) for the use made of abbreviations in the entries of the dictionary. Is there a list of abbreviations in the front matter? What range of information do the abbreviations cover? Can you spot any abbreviations used in the entries which are not included in the list?

2. In what order does the information come in the entries in your dictionary? What order do the senses come in? Check, for example, on the lexemes *bureau, find, precise, tolerate.*

3. Construct a dictionary entry for the verb *regard* from the following citation file:
   (a) We stopped in front of the picture and regarded it for a long time.
   (b) They regard themselves as indispensable to the club.
   (c) We do not regard such matters as our concern.
   (d) They regard what he has done for the club very highly.
   (e) He regarded his sister with disdain.

4. Construct a dictionary entry for the noun *egg* from the following citation file:
   (a) We had eggs for breakfast.
   (b) The baby crocodiles hatched out of their eggs yesterday.
   (c) Our plans are at the egg stage.
   (d) He's a bad egg.
   (e) You shouldn't put all your eggs in one basket.

CHAPTER 16

Lexicology, Lexicography and Semantics

In discussing the meaning of words and how this may be described we have used the three terms in the title of this chapter without a detailed explanation of the differences between them. That is the topic of this final chapter: what lexicology, lexicography and semantics are, and how they each contribute to the description of words and their meaning.

Defining terms

Let us begin with the definitions of these terms as they are given in two dictionaries that we have referred to extensively throughout this book: the *Longman Concise English Dictionary* (*LCED*) and the *Collins English Dictionary* (*CED*).

[1] **lexicology**

> *LCED*: 'a branch of linguistics concerned with the meaning and use of words'
>
> *CED*: 'the study of the overall structure and history of the vocabulary of a language'

The focus of these definitions is quite different. In the *CED* definition the focus is on the whole vocabulary as a mass, whereas in *LCED* the focus is on the parts, individual words. *CED* specifically mentions an historical aspect to lexicology, which is absent from the *LCED* definition; while *LCED* specifically mentions meaning and *CED* does not. The two definitions taken together go some way, as we shall see, towards characterising the term 'lexicology'.

[2] **semantics**

> *LCED*: '**1** the branch of linguistics concerned with meaning **2** a branch of semiotics

dealing with the relation between signs and the objects they refer to'

CED: '**1** the branch of linguistics that deals with the study of meaning, changes in meaning, and the principles that govern the relationship between sentences or words and their meanings **2** the study of the relationships between signs and symbols and what they represent **3** *Logic* **a** the study of interpretations of a formal theory **b** the study of the relationship between the structure of a theory and its subject matter **c** (of a formal theory) the principles that determine the truth or falsehood of sentences within the theory, and the references of its terms'

Here the definitions are similar in the two dictionaries: the second sense in *CED* is the semiotics sense, explicitly referred to by that term in *LCED*. But *CED* adds a third sense which is restricted to the domain of logic. Semantics has a broader scope than just language, having a role in semiotics, the study of signs, of which linguistics (as the study of linguistic signs) is sometimes considered to be a branch. Indeed, as we shall see, there appear to be no clear boundaries to semantics, either in its narrow linguistic sense or in its broader interpretation.

[3] **lexicography**

LCED: '(the principles of) the editing or making of a dictionary'

CED: 'the process or profession of writing or compiling dictionaries'

Both definitions point to lexicography as being concerned with (the process of) making dictionaries. *CED* additionally indicates the use of the term to refer to the profession of those who make dictionaries, while *LCED* ignores this use but sees lexicography as having two parts: a theoretical part, which deals with 'principles'; and a practical part, which deals with the editing process. Again, a conflation of the two definitions would begin to characterise adequately the term 'lexicography'.

Now that we have some general idea of the reference

of these three terms, let us explore each of them in more detail and see how they interact with each other.

Lexicology

Lexicology is a 'branch of linguistics' (*LCED* definition). According to Stephen Ullmann (1962, p. 29), 'it forms next to phonology, the second basic division of linguistic science' (the third is syntax). Lexicology deals with 'significant units', i.e. 'words and word-forming morphemes'. Consequently, there are two basic divisions of lexicology: 'morphology', which studies the forms of words; and 'semantics', which studies the meanings of words. Ullmann thus places semantics within the scope of lexicology: 'This, then, is the place of semantics, in the strict sense of the term, within the system of linguistic disciplines' (ibid.). Many linguists would disagree with this relatively narrow view of semantics, as we shall discuss below. Additionally, Ullmann includes 'etymology' within the scope of lexicology, as the study of the whole history of words, not just of their origins.

Such a view of lexicology encompasses much of what is implied by the dictionary definitions at [1], except that the focus of Ullmann's view is on the study of words as individual units, rather than on the study of the 'overall structure . . . of the vocabulary' (*CED* definition). Lexicology must include both aspects, as contained in the characterisation given by Witold Doroszewski (1973, p. 36):

[4] lexicology is that branch of linguistics investigating words as regards their meaning and use; the science of vocabulary; the theoretical scientific basis of lexicography.

If Ullmann draws a connection between lexicology and semantics, Doroszewski makes the link with lexicography: lexicology represents the theoretical underpinning of lexicography. While that may in theory be the case, lexicography has long had, as we shall discuss below, a tradition that is quite independent of general linguistics, including lexicology. Nevertheless, any lexicographical practice presupposes at least some implicit lexicological theory, since a lexicographer must operate with some notion of what a word (as lexical item) is and what is included in the description and definition of a word. Doroszewski indeed considers 'in a certain sense lexicography . . . a superior discipline to lexicology, for results are more important than intentions, and the value of theoretical intentions must be estimated according to results' (ibid.). That is to say, lexicological theories are useful only in so far as they work, and they are seen to work (or not) in the practical lexicographic description of words.

It will be evident that much of our discussion about words in this book falls within the scope of lexicology as we have outlined it above. Lexicology needs to establish, as we did in Chapter 1, the nature of the phenomena that it studies: what constitutes a lexical item (or lexeme), on the one hand excluding grammatical variants, and on the other including multi-word items (like phrasal verbs) which constitute a single lexeme, as well as distinguishing homonyms and providing criteria for deciding between homonymy and polysemy. Extending the study of the forms of words and including an etymological perspective, lexicology investigates the birth (and death) of words – the topic of Chapter 2: seeing how new words are added to the vocabulary by borrowing or by compounding and derivation, and noting how words become archaic and obsolete. Such an investigation cannot be confined to an examination of word-forms only, but must include a consideration of meaning as well. The etymology division of lexicology studies the origins and history of the forms and meanings of lexemes.

This aspect of lexicology has been approached in this book exclusively in terms of the study of individual lexemes, and this is reflected in the descriptions of the history of the forms and meanings of words in the *Oxford English Dictionary* and in etymological dictionaries. We have not considered an approach which encompasses the overall structure of the vocabulary, nothing that might answer the question, 'What did the vocabulary of English look like in 1400 and how different was it in 1500?'. Such questions may be legitimately included within the scope of lexicology, and they find a practical answer in the *Chronological English Dictionary* (compiled by Thomas Finkenstaedt, Ernst Leisi and Dieter Wolff), an interesting and relatively early use of computing in a lexicographic venture. Finkenstaedt and his colleagues took the data available in the two-volume *Shorter Oxford English Dictionary* and sorted it by computer so that a history of the states of the vocabulary of English was produced. An *Historical Thesaurus of the English Language* (ed. M. L. Samuels) is in an advanced state of preparation at the University of Glasgow. Again computational methods have been adopted, though rather belatedly.

We have paid some attention to this aspect of lexicology in respect of the contemporary language, especially in the discussion of lexical/semantic fields in Chapter 14. In lexical field theory, lexicologists are in fact concerned not only with the overall structure of vocabulary, but also with the description of individual lexemes. Questions such as, 'How many lexical fields are there and how do they inter-relate?', are pertinent to the investigation of the structure of the vocabulary. So are questions that relate to the refer-ential function of vocabulary, such as, 'How does the struc-ture of the vocabulary reflect the way in which the language carves up reality?' But this question is equally applicable to the other concern of lexical field theory: the internal struc-ture of the lexical fields – investigating what lexemes exist for a particular semantic field, how they interrelate, and how they differ in meaning from each other. Lexical field theory is conceived both as a way of making sense of the vocabulary as a whole, and as a way of describing the mean-ings of particular lexemes. This is an area of lexicological theory that could be developed considerably, but the advances may well come through its practical expression in thematic lexicography (see below). Doroszewski (cited above) would not find this surprising.

We can follow on from this by pointing out that our discussion of the external and internal meaning relations of words in Chapters 4, 5 and 6 may also be considered part of lexicology. The external relation of reference or deno-tation, and the internal relations of synonymy, antonymy and hyponymy, which contribute to the description of the meanings of lexemes, are a legitimate concern of the lexi-cologist. The consideration of them belongs to the 'seman-tics' division of Ullmann's characterisation of lexicology, though it is doubtful whether an account of the forms of lexemes and one of their meanings can for many purposes be usefully separated. When we are establishing homonyms, for example, we have to take account of their meanings; equally when we are considering how a lexical field carves up a particular aspect of reality we consider the individual lexemes both as forms to be differentiated and as referring to some bit of reality or experience.

Meaning relations of these kinds belong to the **paradigmatic** axis of language description, the vertical axis, which views linguistic units (in this case, lexemes) as individual items in a substitution relation. Meaning relations on the **syntagmatic** axis, as we noted in Chapter 7, p. 96, also contribute to the description of the meaning of a lexeme. Syntagmatic lexical relations are usually considered in terms of collocation, the mutual expectancy of lexemes. This is another area of lexicology that awaits development, which is now beginning to appear possible with the power and storage capacity of modern computers. A number of attempts to make collocational descriptions using a computer are described by Susan Hockey in *Computer Applications in the Humanities* (1980, pp. 85–91). It is interesting that collocational studies, though they are concerned with meaning, are not usually located within semantics; they are referred to rather as 'lexical' studies, perhaps because the focus is on the cooccurrence of lexical items as items rather than on their referential function as such.

Summarising, lexicology is the branch of linguistics concerned with the study of words as individual items. It thus differs from grammar, since the latter is by and large concerned with words only as members of classes entering into structures; though there is a measure of overlap, as we have seen, in the case of the so-called 'grammatical' words (like determiners). Lexicology deals with both formal and semantic aspects of words: and although it is concerned predominantly with the description of lexemes individually, it also gives attention to a vocabulary in its totality.

Semantics

If you go into a library in which the books are catalogued according to the Dewey Decimal System, you will find books on semantics shelved in two different places. In the Birmingham Polytechnic library, for example, Stephen Ullmann's *Semantics* has the class-mark '149.94', while Frank Palmer's *Semantics* has the class-mark '412'. Ullmann is located among the philosophy books, while Palmer is on the linguistics shelves. Both philosophy and linguistics lay claim to semantics, though it must be said that both Ullmann and Palmer belong to linguistic semantics rather than philosophical semantics. In philosophy it is particularly the area of symbolic logic that has an aspect called semantics, as the definition from *CED* at [2] indicates (sense 3). In linguistics, semantics is sometimes viewed as a 'branch of linguistics' (see definitions at [2]) like lexicology, or phonology, or syntax. Under this view, language is composed of sounds, grammar and meaning; and semantics is the study of meaning.

This view does not have universal assent, however. Frank Palmer, for example, concludes his *Semantics* (1981, p. 206) with the statement that:

[5] . . semantics is not a single well-integrated discipline. It is not a clearly defined level of linguistics, not even comparable to phonology or grammar. Rather it is a set of studies of the use of language in relation to many different aspects of experience, to linguistic and non-linguistic context, to participants in discourse, to their knowledge and experience, to the conditions under which a particular bit of language is appropriate.

With the more recent recognition of a field of 'pragmatics', as the investigation of language in use, Palmer's characterisation of semantics may be seen as including the concerns of pragmatics (see below). There is a sense in which meaning pervades the whole of language, from sounds, where phonemes are established on the basis of meaning differences and where the meaning associations of a sound sequence like /sl-/ or /-mp/ may be investigated, to the meaning of morphemes and words in sentences and the meaning of sequences of sentences in text and discourse, as well as the relations between text and situational context. In the same way that Ullmann sees semantics as a division of lexicology, so it must be seen as a division of phonology, of grammar, of discourse/text linguistics, and of pragmatics.

The isolating of semantics as the 'science of meaning' (the subtitle of Ullmann's *Semantics*), a branch of linguistic science, implies the ability to focus on this pervasive aspect of language to the exclusion of the formal and structural aspects.

If we are to talk about semantics at all therefore, we shall have to identify different kinds of semantics. We can identify, for example, a **pragmatic semantics**, which studies the meaning of utterances in context: e.g. in terms of **speech acts**. We can identify a sentence semantics, which studies the meaning of sentences and meaning relations between sentences: this is sometimes described in terms of truth conditions and implicatures, along the lines of the semantics of formal logic. The kind of semantics that we have been concerned with in this book is **lexical semantics**, which is the title of a recent book on this topic by D. A. Cruse (1986).

Cruse begins the Preface of his book with the comment:

[6] The title of this volume [i.e. *Lexical Semantics*] may lead some readers to expect a book about semantics. They will, I am afraid, be disappointed: the book is, in fact, about the meaning of words. It is not therefore *about* semantics, it is an exercise *in* semantics.

Cruse takes it that the general understanding of semantics is that it is concerned with formal theories of meaning, rather than with the description of word meanings. In a sense, Cruse is putting semantics back to the place allotted it by Ullmann: as a division of lexicology; though Cruse's book pays no attention to the historical dimension of lexical semantics – the development of the meanings of words. The parts of this present book that fall within the scope of Cruse's lexical semantics are Chapter 5 (the 'sense' relations of synonymy and antonymy), the componential analysis of Chapter 6 to some extent, and the discussion of the hierarchical relation of hyponymy in Chapter 14. It is restricted, therefore, to the meaning relations that are internal to the vocabulary of a language. The external relation of denotation may also be regarded as an aspect of lexical meaning and therefore fall within the scope of lexical semantics (Chapter 4).

Summarising, there exists a philosophical semantics and a linguistic semantics, with many shared terms and notions. Philosophical semantics is concerned with the logical properties of language and with the nature of formal theories and the language of logic. Linguistic semantics is concerned with all aspects of meaning in natural languages, from the meaning of utterances in context to the meanings of sounds in syllables. One branch of linguistic semantics is lexical semantics, which studies meaning in relation to words, including both the meaning relations that words contract with each other and the meaning relations that words have with extra-linguistic reality. There seems to be no reason why we should not include within lexical semantics the study of meaning in relation to lexical fields and to collocation. In this sense, lexical semantics may be considered a division of the branch of linguistics we have called lexicology.

Lexicography

If we return to the definitions of *lexicography* at [3], we can identify three ways in which the word is used, though all three uses are clearly related. *Lexicography* is used to refer to a profession, as in the title of R. Ilson (ed.), *Lexicography, an Emerging International Profession* (1986). Here the focus is on the training, the job specification, and the career structure of the relatively small group of professional lexicographers. Many of the papers in Ilson's collection from the 1984 Fulbright Colloquium are concerned with just such matters. We have not discussed lexicography in this sense in this book, though implications for the profession of lexicography might be drawn from some of the discussion in Chapter 15.

A predominant concern in this book has been with a second sense of *lexicography* deducible from the definitions at [3], namely the principles that underlie the process of compiling and editing a dictionary. We may regard such principles as constituting a theory of lexical description, to include both the description of the vocabulary as a whole, and the description of lexemes individually. We traced in Chapter 8 the development of some of the principles of dictionary making in an evolving tradition; for example, the principle of comprehensiveness of coverage which governs what is included in a dictionary, or the historical principle as a way of organising and presenting the lexical description of words. In Chapter 9 we looked at traditions or principles of defining: analytical definition, synthetic definition, encyclopaedic definition, and so on. And in Chapter 15 we discussed some of the decisions that lexicographers have to make in compiling a dictionary and the principles which inform those decisions.

What we have noticed is that some of the principles involved in the making of dictionaries are clearly of a lexical or lexicological nature, while others derive rather from the area of book production. On occasions a decision may be affected by both kinds of principle, and one may be ignored in favour of the other. A decision of this kind is the one that relates to the treatment of lexemes with multiple word-class membership (such as *skin* n., v.). If these are accorded separate headwords because the layout of the page is thereby rendered more attractive, then the decision is informed by the principles of book production. If, however, only one headword is entered, with the consequently longer and denser entry, then it is likely that the decision has been taken on lexical grounds. It is possible, of course, that the separate headwords decision was based on lexical principles also.

To the extent that decisions in dictionary compilation are informed by lexical principles, we may say, as Doroszewski does, that they are derived from lexicological theory. Indeed many aspects of lexicography must derive from explicit or implicit lexicological theory. For example, the question of what constitutes a lexeme is a lexicological matter, including the definition of the class of compounds or the classes of derivations. Lexicology is likewise concerned to investigate questions of homonymy and polysemy, which are of great importance to lexicography. Indeed generally, lexicology investigates how to describe lexemes, both formally and semantically. We have seen in many chapters how lexicological concerns and theories (e.g. componential analysis in Chapter 6) are reflected in the way that words are described in dictionaries. Some lexicological theory (e.g. lexical field analysis) which we may consider of particular relevance to lexicography, has not yet been applied widely in dictionary compilation. This may be either because lexicography as a profession does not or cannot conceive of dictionaries handling lexical description in that way, or because lexicography does not explicitly recognise lexicology as its theoretical basis.

It is probably fair to say that lexicography developed its own principles and traditions independently of the linguistic sciences generally; and it is only in the relatively recent past that explicit links between lexicography and linguistics have been recognised. *Webster's Third New International Dictionary* (1961) was the first to acknowledge the influence of modern linguistics, and then really in two areas only: the representation of pronunciation, and a generally descriptivist rather than prescriptivist stance. Many current dictionaries are, of course, linguistically informed, and compiled by lexicographers who have been trained in linguistics. Indeed, it is not just lexicology which provides descriptive apparatus for lexicography, but other branches of linguistics as well. For example, the study of language variety, which is part of sociolinguistics, contributes to the marking of style and register/domain in dictionaries.

Even so, the old lexicographical tradition still persists in some areas in many current dictionaries. This can be illustrated from the way in which lexemes are marked for word-class. The part-of-speech labels derive from a traditional grammar that was assumed to be common knowledge for all dictionary users, e.g. noun, adjective, verb (intransitive, transitive and absolute); and these traditions and assumptions persist. Besides questioning the usefulness or usability of such labels for modern dictionary users, we may note that developments in grammatical description have been ignored by many lexicographers. This applies both to the recategorisation of items like *this* and *my* as determiners rather than adjectives and the underdifferentiation of the adverb class, as well as to the clear inadequacy of subclasses like 'transitive' and 'intransitive' for verbs and the lack of any useful subclassification for other word-classes, not to mention the sheer incomprehensibility of the term 'absolute' to most modern dictionary users (compare the discussion in Chapter 10, p. 146).

This tension between a persistent tradition and a

realignment with the disciplines of linguistics may also account for the lack of much development in thematic lexicography, which would be the practical descriptive outcome of lexical field theory in lexicology. Certainly, as lexicography takes its appropriate place among the applied linguistic disciplines, we may expect a greater influence in both directions, both from the theoretical to the practical, and from the practical to the theoretical. The practical task of lexical description, which constitutes the essential concern of lexicography, has as much to contribute to the theoretical investigations of lexicology and the other linguistic disciplines as they have to it, since it is above all in practical description that theories will be tested and refined and our knowledge of how language works be increased.

We have mentioned so far two of the uses of the word *lexicography* that can be read from the definitions at [3]. The third refers to the actual process of editing and compiling a dictionary, which was part of the topic of Chapter 15. Clearly, the process of compilation itself and the principles which inform the process cannot be easily separated, and in this book the process of compilation has been of concern only in so far as it reflects the principles of lexicography. We have made no mention of much of the book-production aspect of dictionary making. It is instructive to realise, however, that when you take a dictionary in your hands to consult it, you are holding the product of a vast store of accumulated knowledge and expertise, the present stage of an evolving tradition, moulded by all manner of linguistic and non-linguistic influences. It is a tradition that will go on evolving, as modern linguistics continues to influence the theoretical basis of lexicography and as modern computer technology influences the production possibilities of dictionaries.

Exercises

1. What *does* 'absolute' mean when it is used as a subclassification of verbs in dictionaries (e.g. the *Concise Oxford Dictionary*)? You may find out by consulting a dictionary or grammar book.

2. As a final exercise, take the dictionary that you usually consult and write a critical review of it. You should include in your review a consideration of:

(a) the stated aims of the dictionary and how they are fulfilled

(b) the coverage, in terms of lexemes included, and inclusion policy generally

(c) the kinds of information included for each headword and how they are presented

(d) how the dictionary deals with homonyms, derivations, multiple word-class membership, idioms, etc.

(e) the nature of the definitions, e.g. encyclopaedic, analytical, lucid, etc.

(f) the front-matter and the guidance given to the user

(g) any appendixes, or are abbreviations, prefixes/suffixes, and foreign words and phrases in the body of the dictionary?

(h) any other features of note, e.g. inclusion of biographical or geographical information

Key to Exercises

Chapter 1. What is a Word?

1. *sea* is polysemous and has the homophone *see*
break is polysemous and has the homophone *brake*
line has three homonyms, each of which is polysemous
ear has two homonyms, one of which (= 'organ of hearing') is polysemous
prayer has a homograph (*pray* + *er* = 'one who prays'), and is itself polysemous
mature has two homonyms, each of which is polysemous
trace has three homonyms, two of which are polysemous
house has a homograph (i.e. the verb) and is polysemous

You are encouraged to consult a dictionary for full details of these words.

2. child   child's   children   children's
run   runs   running   ran
little   littler   littlest;   **also** less   lesser   least
fly   flies   flying   flew   flown
basic   more/most basic
turn (n.)   turns
turn (v.)   turns   turning   turned

3. *Lexemes*: 16 (counting *drum roll* and *roll out* each as one lexeme)
Word-forms: 18 (counting *drum, roll* (n.), *roll* (v.), *out* as separate forms, but not *have to*)

Orthographic words: 18 (counting *roll* (n.) and *roll* (v.) as one word, and *have* and *to* as two words)

4. Multi-word lexemes are:
   take care of (compare *nurse*)
   look into (prepositional verb, compare *investigate*)
   story book (compound)
   garden fence (possibly, though compound status not so well established as *story book*)
   send off for (phrasal prepositional verb, compare *request*)
   over the moon (idiom, non-literal meaning)
   training weekend (compound)
   look up (phrasal verb)

5. *Lexical words*: am   grown   man   estate   be   proud   great   tell   other   girls   boys   to meddle   toys
   Grammatical words: when (conjunction)   I (pronoun)   to (preposition)   shall (modal auxiliary verb)   very (intensifying adverb – possibly lexical)   and (conj.)   the (determiner)   not (negative particle)   with (prep.)   my (possessive determiner)

Chapter 2. Where did English Words Come From?

1. The Anglo-Saxon words are: let   lewd   liar   lick   lid   life.
   The others derive mostly from French, one directly from Latin (*ligament*).

2. The French loan-words are: pedal   pedometer   peignoir   pellet   pencil   pension   perform   perfume.
   The others are Anglo-Saxon words.

3. The Latin loan-words are: subdivide   subsidy   suburb   suction   suffix   suggest   superb.
   The others are either Anglo-Saxon or loans from French (*suede, sugar*).

4.   cheap – inexpensive
      cheeky – insolent, impudent
      hard – impenetrable

lighting – illumination
busy – occupied, engaged
buy – purchase
worker – employee
cross – traverse
own – possess
give – donate

5. *LCED* lists the following:
   geocentric geochronology geology geomagnetic
   geomorphology geophysics geostationary geos-
   trophic geosynchronous geothermal geotropism.
6. (a) One might suggest, for example, *tablement* (by
       analogy with *statement*), *defication* (by analogy with
       vilification).
   (b) One might suggest *minutely* (by analogy with *hourly*,
       daily), and *widowish* (by analogy with *girlish*).
   (c) One might suggest *worsement* (by analogy with
       betterment), and *see-throughency* (by analogy with
       transparency).

The degree of success you might have in persuading
your fellow-speakers of the desirability of these neol-
ogisms is a matter of conjecture.

Chapter 3. Dictionaries: the Repositories of Words

1. *agate* has alternative pronunciations of the second
   syllable, either ['ægət] or ['ægeit]
   chaffeur is a loan-word from French; some speakers
   retain a French-sounding ending [ʃəʊ'fɜ:], other anglicise
   it ['ʃəʊfə]
   dimension has alternative pronunciations of the first
   syllable, either [dɪ'mɛnʃən] or [daɪ'mɛnʃən]
   either has alternative pronunciations of the first syllable:
   ['aɪðə] or ['i:ðə]
   lichen has alternative pronunciations: ['laɪkən], ['litʃən]
   longitude has alternative pronunciations: ['lɒndʒɪ,tʃu:d],
   ['lɒŋgɪ,tʃu:d]; the final syllable is pronounced alternatively [tju:d]
   paella is a loan-word from Spanish; its pronunciation is

usually anglicised [paɪ'ɛlə], but you may retain a
Spanish-sounding pronunciation [pa'elə]
punctuate has alternative pronunciations of *-nctu-*:
['pʌŋktjʊ,eɪt], ['pʌŋktʃʊ,eɪt], ['pʌntjʊ,eɪt]
strength has alternative pronunciations of the final consonant cluster: [strɛŋkθ], [strɛŋgθ], [strɛnθ]
Uranus has alternative stress patterning together with
variant vowels: [jʊ'reɪnəs], ['jurənəs]

2. Comparing *CED* with *LCED*: both have separate entries
   for *calculator, encourage, preeminent, rusty*; both have run-
   ons for *heaviness, musicologist*. For *flattish, LCED* has a
   run-on, while *CED* has a separate entry; similarly for
   survivor and *vaccination*. For *graceless, LCED* has a
   separate entry, while *CED* has a run on.

3. *CED* has the following usage labels:
brass (= 'money')	Northern English dialect
caddy	Golf
depreciable	U.S. (i.e. American English)
featly	Archaic
heebie-jeebies	Slang
j'ouvert	Chiefly Caribbean
maggoty (= 'annoyed')	Austral. Slang
once-over	Informal
ritenuto	Music
titfer	Cockney rhyming slang

4. In *LCED* there are four headwords *cock* (i.e. four
   homonyms/homographs), in *CED* only two. *1cock* in
   LCED is a noun; six senses are distinguished and sense
   1 has two subsenses. *2cock* is a verb; it has three senses,
   and senses 1 and 2 each have two subsenses; the entry
   contains additionally the idiomatic expression *cock a
   snook*. These two headwords are combined into *cock1* in
   CED probably because of a common etymology. *3cock*
   in *LCED* is a noun with only one sense; and *4cock* is the
   related verb: these are combined into *cock2* in *CED*,
   again with a common etymology, which is different
   from that of *cock1*.

5. In *CED*, 'the first sense given is the one most common

in current usage' (p. xv); alternatively a 'core meaning' may be placed first. General senses precede technical senses, followed by archaic and obsolete senses, and then idiomatic and fixed expressions.

In *LCED*, 'those meanings that would be understood anywhere in the English-speaking world are shown first in their historical order'. They are followed by meanings with restricted usage or no longer in current use.

Chapter 4. Words and the World

1. The *LCED* definitions of the items are as follows:
   cup: 'a small open drinking vessel that is usu bowl-shaped and has a handle on 1 side'
   jam: 'preserve made by boiling fruit and sugar to a thick consistency'
   path: 'a track formed by the frequent passage of people or animals; a track specially constructed for a particular use'
   screw: 'a usu pointed tapering metal rod having a raised thread along all or part of its length and a usu slotted head which may be driven into a body by rotating (e.g. with a screwdriver)'
   wine: 'fermented grape juice containing varying percentages of alcohol together with ethers and esters that give it bouquet and flavour'

2. (a) employ   hire   engage   appoint   take on   recruit   staff   fill (a post)
   (b) dismiss   fire   discharge   make redundant   sack   lay off   retire

3. (a) aromatic   fragrant   perfumed   scented   sweet-smelling
   (b) acrid   evil-smelling   fetid   fusty   musty   noisome   pungent   putrid   rank   smelly   stinking

4. *charity*: religious; or possibly pejorative, as something underserved or dispensed by the privileged/authority
   iron: possibly strength, durability; or capacity to rust
   mole: pleasant, soft; or pest; or unpleasant, unobservedly subversive

snow: pleasant, play, winter sports; or unpleasant, nuisance, travel difficulties

street: danger; or play, friendship, street life

5. All of these words may be defined 'scientifically'. *CED* defines them all in this way. *LCED* defines *sodium* and *spider* in this way; *mushroom* and *robin* have a scientifically oriented definition; and *badger* is least scientific in its definition.

Chapter 5. Words and Words

1. *Chamber* (from French for 'room') and *room* have taken on different meanings; they are now partial synonyms. *Chamber* refers to a large room used for official or institutional purposes.

   Fleer has become obsolete. So have *reck* and *sooth* (except in the phrase 'in sooth' and *soothsayer*).

   Spirit and *ghost* have taken on different meanings: *ghost* now generally refers to apparitions and *spirit* to the non-corporeal part of human beings, or abstractly to the atmosphere, motivation, etc. of an action or idea.

   Judgement has taken over as the word with general reference; *doom* has become specialised in meaning in collocations like 'going to one's doom' (i.e. unavoidable death or destruction), or in fixed expressions like 'full of gloom and doom'.

2. *bale* and *bundle* are partial synonyms: they tend to be applied to different materials

   cicatrix is technical for *scar*

   depression is more formal style than *slump*

   gowk is dialect for *fool*

   lumber is technical and North American for *logs*

   naturism is a euphemism for *nudism*

   remuneration is formal style for *pay*

   sufficient is formal style for *enough*

   teem has connotations of *abounding* in very large numbers

   umbilicus is technical for *navel*

3. The gradable antonyms are: fast – slow   high – low   rich – poor   thin – fat.

The complementary antonyms are: captive – free fixed – loose in – out leave – stay.
The converses/relational opposites are: behind – in front north of – south of parent – child teacher – pupil.

4. *Banjo, beaker* and *blancmange* do not use synonymy in their definitions in *LCED*. The remainder do, either exclusively, or in part.

Chapter 6. Analysing Word Meanings

1. The common semantic components are [+ANIMAL], [+DOMESTIC]. To distinguish cats and dogs we need either the component [FELINE] or the component [CANINE]; to distinguish among a larger set of domestic animals we would need both.

	[FELINE]	[ADULT]	[MALE]
bitch	–	+	–
cat	+	+	–/0
dog	–	+	+/0
kitten	+	–	0
puppy	–	–	0
tomcat	+	+	+

Note that neither of the young are marked for gender; and in the case of the adults, *dog* is male or general, while *cat* is female or general.

2. The common semantic component is [DRINKING VESSEL], or perhaps these should be two components. The crucial discriminators seem to be: what the items are made of, whether they have a handle, what they are used for, and to distinguish *cup* and *mug* whether they are used with a saucer.

	[MATERIAL]	[HANDLE]	[USE]	[SAUCER]
cup	ceramic	+	hot liquid	+
goblet	glass	–	wine	–
mug	ceramic	+	hot liquid	–
tankard	glass/metal	+	beer	–
tumbler	glass	–	cold liquid	–

W. Labov has argued that componential analysis cannot be applied to items like *cup*, because their attributes are 'fuzzy' in meaning. But Geoffrey Leech has defended the use of componential analysis for such items on the basis that it is a 'prototypical' cup that we have in mind. Reference to the debate can be found in Leech's *Semantics* (1981, p. 121f.).

3. The common semantic component is [MUSICAL INSTRUMENT]. Their meanings can be discriminated by: material from which they are constructed, method of playing.

	[MATERIAL]	[METHOD]
clarinet	wood	blow
cymbal	metal	strike
harp	wood/gut	pluck
trumpet	metal	blow
violin	wood/gut	scrape

It will be noted that [METHOD] deviates from the traditional division of instruments of the orchestra into: strings, woodwind, brass and percussion. But these two features suffice to make the discriminations, though by no means all has been said about the meanings of these lexemes that could be said.

4. In the definitions of these lexemes in *LCED*, those for *deprive, litotes, nonchalant* and *tortuous* appear not to be informed by (implicit) componential analysis. The definitions of the other lexemes do to some extent. In most cases the lexeme is related to a general set of lexemes (e.g. rondeau, 'a form of verse') and then the characteristic features are given (e.g. 'using only two rhymes', 'opening words of first line used as a refrain'), implying a discrimination of the meaning of this lexeme from that of related lexemes (i.e. other kinds of verse).

Chapter 7. Meaning from Combinations

1. *Collocations*: c, f, h
   Proverbs: e, i
   Idioms: a, d, g
   Simile: b

2. *accuse*: persons of a supposed crime or misdemeanour
   betray: friend, one's country
   put on: clothes, weight
   repair: machine, model or other artefact
   sing: song, madrigal, cantata, motet, etc.
   utter: sound, speech

3. (a) speak frankly
   (b) finish
   (c) fail, be unsuccessful
   (d) do something thoroughly
   (e) in debt
   (f) completely
   (g) behave properly
   (h) agree with
   (i) ostentatiously

4. The *LCED* entry for *drive* contains a number of speci-
   fications of possible objects for various senses of the
   verb, e.g. 'to control and direct the course of (a vehicle
   or draught animal)', 'to bore (e.g. a tunnel or passage)',
   'to propel (an object of play) swiftly'. It also includes the
   phrasal verb *drive at*, and the idiom 'drive up the wall'
   (= 'infuriate', 'madden').

Chapter 8. Why Dictionaries?

1. The answer to this question is more easily deduced from
   the *Plan*, where he discusses in order: orthography (i.e.
   spelling), pronunciation, etymology (or derivation),
   analogy (i.e. inflections), syntax (e.g. verb +
   preposition), phraseology (i.e. collocation), interpretation (i.e. the definition), and distribution (i.e. restrictions on use, such as 'poetical', 'obsolete'). These are all,

apart from phraseology and distribution, mentioned in the Preface: orthography, pronunciation and etymology specifically; analogy and syntax by the phrase 'words grammatically considered'; and interpretation is renamed 'explanation' or 'signification'. Additionally, the lexicographic description will be complemented by quotations or examples.

2. (a) Main words constitute the bulk of the vocabulary and the other two types are defined in relation to them. Subordinate words comprise: obsolete or variant forms of main words; irregular or peculiar inflections of main words; words of 'bad formation' or doubtful existence. Combinations involve more than one main word in a fixed phrase or collocation.

(b) These are listed as: Identification (including spelling, pronunciation, part-of-speech, special use (e.g. Music), status (e.g. obsolete), earlier forms or spellings, irregular inflections); Morphology, i.e. the history of forms; Signification, i.e. explanation of meanings; Quotations.

(c) These terms refer to the origins of words. 'Naturals' refers to native and 'fully naturalised' words. 'Denizens' are naturalised in use but not in form, inflection or pronunciation (e.g. *aide-de-camp*). 'Aliens' are words referring to foreign objects for which there is no native term (e.g. *kibbutz*). 'Casuals' are foreign words of restricted perhaps temporary use (e.g. *Ostpolitik*, used of the policy of detente towards East Europe pursued by West Germany in the 1970s; or *sputnik*, used only of Russian space satellites).

3. (a) There is a new 'pronunciation alphabet', i.e. transcription system, which is 'designed to represent clearly the standard speech of educated Americans'. On the other hand the dictionary claims to show 'the pronunciations prevailing in general cultivated conversational usage, both informal and formal, throughout the English speaking world', but it does not 'attempt to dictate what that usage should be'.

(b) *Webster's Third* claims to be more consistent in its

etymological information than previous editions. Four general kinds of etymology are recognised: native words, old and well-established borrowings, more recent borrowings, and borrowings from non Indo-European languages. Compare the answer to 2(c) for *OED*. The label 'ISV' is introduced, standing for 'International Scientific Vocabulary', to account for scientific words of recent coinage that are often composed from Latin or Greek roots.

Chapter 9. How to Define a Word

1. Five senses are arguably identifiable from these examples:
   (1) 'perceive with the eye' (a, e)
   (2) 'inspect' (b)
   (3) 'understand' (c)
   (4) 'be a spectator at' (d)
   (5) 'have a meeting with' (f)
   See has many other senses: look it up in your dictionary.

2. If the senses are historically ordered, the biblical meaning of *scapegoat* will be given first, and then the more general current meaning, i.e. 'a person made to bear the blame for others'.

3. *here*: probably synonyms and rule-based
   milk: all senses probably defined analytically
   humdrum: probably by synonyms
   hundred: probably synthetic
   hypermarket: probably analytic

4. Any of these lexemes may have encyclopaedic defini-tions, if your dictionary tends to use them for words that have been classified scientifically.

Chapter 10. More than Meaning

1. There may not be a definition as such at all for these words, as in *LCED*, which merely notes that they are the 'objective case' of *we* and *they* respectively. *LCED* notes a second sense of *us*, as a substitute for *me*, but labels this 'non-standard'. *Them* is given a homonym,

defined by the synonym 'those' and with the example 'them blokes'; it is labelled 'non-standard'.

2. *Cloth* is a noun, with a regular plural by the spelling (*cloths*), but with alternative pronunciations /klɒθs/ or /klɒðz/.

Have is a verb, with irregular third person singular present (*has*), past tense and past participle (*had*); it occurs as a transitive verb (vt) and as an auxiliary verb (va).

Little occurs as an adjective, with comparative forms *littler, less* and *lesser*, and superlative forms *littlest* and *least*. It occurs also as an adverb, with comparative form *less* and superlative form *least*. It also occurs as a noun.

Memorandum is a noun, with variant plurals *memorandums* (regularised) and *memoranda* (original Latin).

Up occurs as an adverb, adjective, verb, preposition and noun. As a verb it doubles the 'p' when affixed (e.g. *upped*).

3. Examples of usage labels might be (from *CED*):

jail delivery: sense 2 ('a commission issued to assize judges . . .') has domain label 'English law'

jakes: 'archaic' 'slang' word for *lavatory*; southwestern English dialect for human excrement

jammy: 'British slang'

jangle: sense 3, 'archaic' for *wrangle*

jar: sense 3, 'Brit informal' = glass of alcoholic drink, esp. beer

Chapter 11. Different Dictionaries

1. Again the *Concise* entry represents a trimming of the *CED* entry: sense **4** is omitted altogether, as is **2b**, which is deducible from **2a**. The definition of sense **3** is truncated by omitting examples of leguminous plants beyond 'sweet pea'; and that of sense **1** omits the technical term 'papilionaceous'. The derivative *pealike* is omitted. In the etymology the *Concise* has more abbreviation: '<' for 'from', 'pl' for 'plural'. Interestingly, the etymology in the *Pocket* is fuller than in the other two dictionaries, though they in fact have the additional information under the headword *pease*. Senses **1** and **2** of the *Pocket*

correspond with those in *CED* and the *Concise*; the definitions are considerably contracted. The remaining sense(s) are omitted. The *Pocket* entry includes the simile 'as like as two peas (in a pod)'.

2. While providing the origin of the word itself (i.e. from Hindi '*Jagannath*'), the *Collins Pocket* etymology is otherwise rather unrevealing. The *Longman Concise* at least reveals the association of meaning that caused the word to be borrowed into English. *The Oxford Etymological* additionally provides the date of borrowing (seventeenth century), and the Sanskrit original of the Hindi word.

3. We noted in the chapter the omission of certain items of information from the children's dictionaries: pronunciation, etymology, etc. They also make a selection of the senses: *Oxford* excludes the 'evasive talk' meaning, *Longman* the 'trousers' meaning. There is no entry for the verb senses in the *Oxford*, and they take up a separate entry (not given) in the *Longman*. The *Oxford* definitions are clearly deliberately simple, and the *Longman* ones perhaps slightly simpler than the *Collins*. Note that the label 'colloq.' in *Collins* is replaced by 'esp. spoken' in *Longman*. Similarly the abbreviation '[pl.]' of sense **3** of *Collins* is replaced by the plural form in sense **2** of *Oxford*.

Chapter 12. Especially for the Learner

1. Some of the principal differences to note are: (a) the entries are organised differently: *LCED* makes a basic division between transitive (vt) and intransitive (vi) uses, while *LDOCE* divides on meaning between the theatrical and the recounting sense; (b) consequently, the transitive (T) and intransitive (I) theatrical uses are, sensibly, brought together under the same sense; (c) *LCED* distinguishes two recounting senses, where *LDOCE* has one, and these are put first in *LCED* (presumably because they are historically prior), while this sense comes second in *LDOCE* (because it is less common), where it is marked as 'formal'; (d) the *LDOCE* definitions are simpler and do not use the morphologically related form *rehearsal* as *LCED* does;

(e) each of the senses and each syntactic possibility for the 'practise' meaning has an example in *LDOCE*, but not in *LCED*; and sense 2 in *LDOCE* has a cross-reference to *recount*, which does have an example. In a number of ways, therefore, the *LDOCE* entry might be regarded as a superior lexicographic description to that in the *LCED*.

2. According to *LDOCE*, *intend* is transitive (T) in all its uses. It may be followed by a *to*-infinitive clause [+ obj + to-v], e.g. 'I intend them to go'; or by a *that*-clause with omissable *that* [+ (that)], e.g. 'I intend (that) they should go'. Additionally, *intend* may enter a pattern with an object and a prepositional phrase introduced by *for* or *as* [+ obj + for, as], though this pattern is usually found in the passive, e.g. 'It was intended as a joke'. *OALD* adds to these: a pattern with noun phrase object, e.g. 'Does he intend marriage?', a pattern with gerund object, e.g. 'What do you intend doing today?'.

3. *LDOCE* recognises three basic senses for the noun *cake;* *COBUILD* recognises two, with two subdivisions of the first sense. Sense 1 of *LDOCE* corresponds to 1.1 of *COBUILD*; sense 2 of *LDOCE* corresponds to senses 1.2 and 2 of *COBUILD*; and sense 3 of *LDOCE* is not given in *COBUILD*. Both dictionaries begin with the obvious central meaning of *cake* and label it both countable and uncountable: *LDOCE* combines the two uses by beginning the definition with '(a piece of)' to indicate the countable use; *COBUILD* defines the countable use first and then shows how it is used as an uncountable noun. *LDOCE* includes a number of cross-references to specific kinds of cake and to a closely related item (biscuit). *COBUILD*'s sense 1 is based on the use of *cake* in reference to food, while sense 2 gives the non-food use in relation to soap and the like. *LDOCE* combines (in its sense 2) the food and non-food use of *cake* in reference to the shape of something. Both note that the use of this sense of cake usually involves further specification: 'often in comb(inations)' – *LDOCE*, 'N COUNT + SUPP(orting word or phrase)' – *COBUILD*. Both entries have a good range of examples.

Chapter 13. Who Uses a Dictionary for What?

2. *aggravate*: the 'irritate' use is disapproved by some
speakers. *Collins Concise* marks this sense 'informal'
and the *Concise Oxford* (Fifth Edition) 'colloquial'.
alright is marked 'Not standard' in *Collins Concise* and
'nonstandard' in *Longman Concise*.
decimate: the literal 'every tenth man' sense is the only
one approved by some people; dictionaries usually
follow general usage and include the 'destroy large
proportion of' sense.
different: the use of *to* after *different* is disapproved by
some speakers. *Longman Concise* notes: '+ *from*, chiefly
Br. *to*, or chiefly NAm *than*'. *Concise Oxford* (Fifth
Edition) notes: '(*from, to, than*, all used by good writers
past and present, *than* chiefly where a prep. is
inconvenient)'.
due to: *Concise Oxford* notes: 'the advl use for "owing",
as "I came late ~ an accident", is incorrect'. *Longman
Concise* notes: '**due to** prep BECAUSE OF **1** – though
disapproved by many, now used by many educated
speakers and writers . . .'
gaol/jail: *Concise Oxford* (Fifth Edition) has a full entry
under *gaol*, but a note 'see GAOL' under *jail*. *Longman
Concise* and *Collins Concise* note *gaol* as a variant British
spelling of *jail*, which has the full entry.
media: *Longman Concise* notes it merely as a 'pl of
MEDIUM'. *Collins Concise* has additionally: '*Inf.* the mass
media collectively'. Under sense **2(b) (2)** of *medium*,
LCED notes that it may be 'pl but sing or pl in
constr[uction]', with the meaning 'MASS MEDIA'.
principal/principle: dictionaries do not explicitly help to
disentangle such commonly confused pairs (cf. also
complement/compliment, lay/lie) except by the way in
which they define the individual lexemes.

3. In the *Longman Concise* the items are listed as:

   anti-slavery: not included
   beautification: run on under *beautify*
   hacker: run on under *6hack* ('to ride (a horse) at an
   ordinary pace, esp over roads'). The new computer
   sense has a separate entry in the Second Edition
   (1986) of *Collins English Dictionary*.

me-too-ism: not included

openness: run on under *1open* adj

prettify: separate entry

privatise: run-on under *privatisation*, implying that it is a back-formation from this. *Collins Concise* has *privatise* as the headword and *privatisation* as the run-on.

randomiser: run on under *randomize*

re-employ: not included; in *Collins Concise* it is among the list of derived words given at the bottom of the page.

tankful: run-on under *tank*

Chapter 14. Not Alphabetical

1. The *Longman Concise English Dictionary* contains the following BUILDING words beginning with *ga-*:

gable   gallery   gambrel   gangway   garage   garde-robe   gargoyle   garner   garret   gate   gatehouse   gazebo

We can make an initial distinction between those referring to kinds of buildings, and those referring to parts of buildings. In the first group would come: *garage garner gatehouse gazebo*. And they would be distinguished according to their use or purpose: keeping vehicles in, storing grain, etc.

From the second group we can distinguish a set referring to different kinds of room: *gallery garderobe garret* – differentiated by shape, purpose and position. A second set refers to different kinds or parts of roofs: *gable gambrel gargoyle*. And a third set refers to aspects of access to or within buildings: *gangway*

2. A feature of 'causing' or 'initiating' an action is contained in the meanings of *send* and *dispatch*, while a feature of 'carrying' is implied in *transport, convey* and *deliver*. *Deliver* also has a feature of 'destination', which is absent from the others. A feature of 'goods' seems present in the meanings of *transport* and *dispatch*, while for the other lexemes 'messages' may also be sent, conveyed or delivered. These features would probably serve to distinguish these five lexemes.

3. These lexemes all refer to the activity of recall or remembering, of thinking into the past and bringing into present focus. In the *Thesaurus* they occur in the set *505 Memory* among a large number of other words. This set finds its place in the subsection 'Extension of Thought: To the Past' of Class IV Intellect.

4. These lexemes refer to 'ways of cutting meat for cooking'. *Cut* is a kind of superordinate term for the whole set. They differ largely in the size, shape and provenance of the cut. For example, *slice* and *rasher* both refer to thin cuts of meat, but we talk of a 'slice of veal' and a 'rasher of bacon'. A *joint* implies a largish cut of meat, usually with a bone, e.g. from the shoulder or leg of an animal. Without a bone, it would be a *fillet*. A *chop* usually comes from the breast of an animal and contains a piece of meat on a bone. A *steak* is a thick cut of meat, and a *cutlet* is a small one.

Chapter 15. The Craft of Lexicography

1. The list of abbreviations will no doubt include general abbreviations like *approx., cf., esp., pl.* and *usu.*, as well as traditional part-of-speech abbreviations like *n., v., prep.* and *conj.* Additionally there may be regional labels like *Am., Brit., dial.*; style labels like *colloq., infml, sl.*; and register labels like *Biol., Chem., Mus., Naut.*

2. A usual order of information after the headword is: pronunciation, part-of-speech, (irregular) inflections, definitions – with dialect, style and register labels before the definition for each sense, and finally etymology. The senses are usually arranged either in historical order or in order of supposed currency or frequency of occurrence; idioms and other fixed expressions usually come at the end of the entry.

3. *LCED* has the following entry for *regard*:

   regard . . . 1 to pay attention to; take into consideration **2** to look steadily at **3** to relate to; concern **4** to

consider and appraise in a specified way or from a specified point of view . . .
Sentence (a) illustrates sense **2**; (b), (c) and (d) illustrate sense **4** (the most frequently occurring in modern English?); and (e) illustrates sense **1**. Sense **3** is probably most often found in the form 'Regarding . . .' or 'With regard to . . .'.

4. *CED* has the following entry for *egg*:

    egg . . . **1** the oval or round reproductive body laid by the females of birds, reptiles, fishes, insects, and some other animals, consisting of a developing embryo, its food store and sometimes jelly or albumen, all surrounded by an outer shell or membrane. **2**. Also called: **egg cell**. any female gamete; ovum. **3**. the egg of the domestic hen used as food. **4**. something resembling an egg, esp. in shape or in being in an early stage of development. **5. good** (or **bad**) **egg**. *Old fashioned informal* a good (or bad) person . . .
    Sentence (a) illustrates sense **3**; (b) illustrates sense **1**; (c) illustrates sense **4**, (d) illustrates sense **5**; and (e) is an idiom which will come later in the entry with the meaning 'to stake everything on a single venture' (*CED*).

Chapter 16. Lexicology, Lexicography and Semantics

1. The term 'absolute' is used for verbs like *read* and *write*, which are normally regarded as transitive, but where the syntactic object may be omitted, to refer to the activity in general, e.g. 'The class is reading' (i.e. engaged in the activity of reading, without specifying what they are reading, contrast 'The class is reading *Lucky Jim*'). The new (eighth) edition of the *COD* will continue to use the term 'absolute'.

2. It would be impossible to provide a satisfactory key to this open-ended exercise. You have, it is to be hoped, found the exercise profitable and been able to use the knowledge that you have gained by studying this book.

Further Reading

The following books are recommended as a follow-up to the topics discussed in this book.

Lexicology and semantics

Cruse, D. A., *Lexical Semantics*. CUP 1986
Hurford, J. R. & Heasley, B., *Semantics, a Coursebook*. CUP 1983
Leech, G., *Semantics* (2nd edn). Penguin 1981
Lyons, J., *Language and Linguistics*. CUP 1981, Ch. 5
Palmer, F. R., *Semantics* (2nd edn). CUP 1981
Ullmann, S., *Semantics*. Basil Blackwell 1962

History of lexicography

McArthur, T., *Worlds of Reference*. CUP 1986
Walker-Read, A., 'Dictionary', in *New Encyclopaedia Britannica* 1974. Macropaedia Vol. 5, pp. 713–22

Lexicography

Hartmann, R. R. K. (ed.), *Dictionaries and their Users*, Exeter Linguistic Studies Vol. 4. Exeter University 1979
Hartmann, R. R. K. (ed.), *Lexicography – Principles and Practice*. Academic Press 1983
Ilson, R. (ed.), *Dictionaries, Lexicography and Language Learning*. Pergamon Press/British Council 1985

Ilson, R. (ed.), *Lexicography – An Emerging International Profession*. Manchester UP/Fulbright Commission 1986

Landau, S. I., *Dictionaries – the Art and Craft of Lexicography*. Charles Scribner's & Sons 1984

Works Referred to

Dictionaries

Cassell's English Dictionary, eds. A. L. Hayward *et al.* 1962
Cassell's Spelling Dictionary (2nd edn), compiled by David Firnberg. 1985
Chambers Twentieth Century Dictionary ed. A. M. Macdonald. 1977
Chronological English Dictionary, eds. T. Finkenstaedt, E. Leisi, D. Wolff. Universitätsverlag Heidelberg 1970
Collins Cobuild English Language Dictionary, ed. J. Sinclair, 1987
Collins English Dictionary, ed. P. Hanks. 1979; 2nd edn 1986
Collins/Klett English–German Dictionary. 1983
Collins Pocket English Dictionary, ed. W. T. McLeod. 1981
Concise Oxford Dictionary (5th edn), ed. E. McIntosh. 1964
Concise Oxford Dictionary of English Place Names (4th edn), ed. E. Ekwall. 1960
Dictionary of Anagrams, Samuel C. Hunter. RKP 1982
Dictionary of Data Processing and Computer Terms, R. G. Anderson. Macdonald & Evans 1982
Dictionary of Foreign Words and Phrases, Alan Bliss. RKP 1966; paperback edn 1983
Dictionary of Slang and Unconventional English (8th edn), Eric Partridge, ed. Paul Beale. RKP 1984
Dictionary of the English Language (4th edn), Samuel Johnson. 1775
Duden Rechtschreibung (17th edn). Bibliographisches Institut Mannheim 1973
Everyman's English Pronouncing Dictionary (13th edn), Daniel Jones, ed. A. C. Gimson. J. M. Dent 1967

Heinemann English Dictionary, eds K. Harber & G. Payton. 1979

Historical Thesaurus of the English Language, ed. M. L. Samuels. In preparation

Longman Concise English Dictionary, ed. P. Proctor. 1985

Longman Dictionary of Contemporary English, ed. P. Proctor. 1978

Longman Dictionary of Contemporary English, New Edition, ed. D. Summers. 1987

Longman Dictionary of English Idioms, eds. T. H. Long & D. Summers. 1979

Longman Dictionary of the English Language, eds H. Gay, B. O'Kill, K. Seed & J. Whitcut. 1984

Longman Lexicon of Contemporary English, Tom McArthur. 1981

Longman New Generation Dictionary, ed. P. Proctor. 1981

New Collins Concise English Dictionary, ed. W. T. McLeod. 1982

New Zealand Pocket Oxford Dictionary, ed. R. W. Burchfield. 1986

Oxford Advanced Learner's Dictionary of Current English (3rd edn), ed. A. S. Hornby, with A. P. Cowie. 1974; revised 1980

Oxford Children's Dictionary, compiled by John Weston & Alan Spooner. Granada Publishing & OUP 1976

Oxford Dictionary of English Etymology, ed. C. T. Onions, with G. W. S. Friedrichsen and R. W. Burchfield. 1966

Oxford English Dictionary, eds. J. A. H. Murray, H. Bradley, W. A. Craigie & C. T. Onions. 1933

Oxford Illustrated Dictionary (2nd edn), eds. J. Coulson *et al.* 1975

Oxford Paperback Dictionary, compiled by Joyce M. Hawkins. 1979

Penguin Rhyming Dictionary, Rosalind Fergusson. Viking 1985

Pocket Oxford Dictionary (7th edn), ed. R. E. Allen. 1984

Roget's Thesaurus of English Words and Phrases (new edn). Longman 1936

Webster's New Collegiate Dictionary, ed. H. B. Wolf. 1977

Webster's Third New International Dictionary, ed. P. Gove. 1961

Other works

Béjoint, H., 'The Foreign Student's Use of Monolingual English Dictionaries', *Applied Linguistics* II/3, 1981, pp. 207–22

Brook, G. L., *Words in Everyday Life*. MacMillan 1981

Burton-Roberts, N., *Analysing Sentences*. Longman 1986

Cruse, D. A. *Lexical Semantics*. CUP 1986

Doroszewski, W., *Elements of Lexicology and Semiotics*. Polish Scientific Publishers/Mouton 1973

Ellegård, A., 'On Dictionaries for Language Learners', *Moderna Språk* LXXII, 1978, pp. 225–44

Fernando, C. & Flavell, R., *On Idiom – Critical Views and Perspectives*, Exeter Linguistic Studies Vol. 5. Exeter University 1981

Fieldhouse, H., *Everyman's Good English Guide*. J. M. Dent & Son 1984

Foster, B., *The Changing English Language*. MacMillan 1968

Hanks, P., 'To What Extent Does a Dictionary Definition Define', in R. R. K. Hartmann (ed.), 1983

Hartmann, R. R. K. (ed.), *Dictionaries and their Users*, Exeter Linguistic Studies Vol. 4. Exeter University 1979

Hartmann, R. R. K. (ed.), *Lexicography – Principles and Practice*. Academic Press 1983

Hockey, S., *Computer Applications in the Humanities* Duckworth 1980

Hornby, A. S., *A Guide to Patterns and Usage in English*. OUP 1954

Hulbert, J. R., *Dictionaries, British and American* (revised edn). Deutsch 1968

Ilson, R. (ed.), *Dictionaries, Lexicography and Language Learning*. Pergamon Press/British Council 1985

Ilson, R. (ed.), *Lexicography, An Emerging International Profession*. Manchester UP/Fulbright Commission 1986

Johnson, S., 'The Plan of a Dictionary of the English Language', in M. Wilson (ed.), *Johnson – Poetry and Prose* (2nd edn). Rupert Hart-Davis 1957

Kipfer, B. A., *Workbook on Lexicography*, Exeter Linguistic Studies Vol. 8. Exeter University 1984

Knowles, G., *Patterns of Spoken English*. Longman 1987

Leech, G. N., *Semantics* (2nd edn). Penguin 1981

Lehrer, A., *Semantic Fields and Lexical Structure*. North Holland Publishing Co. 1974

Lounsbury, F. G., 'The Structural Analysis of Kinship Semantics,' in H. G. Lunt (ed.), *Proceedings of the Ninth International Congress of Linguists*. Mouton 1964

Lyons, J., *Language and Linguistics*. CUP 1981

McArthur, T., *Worlds of Reference*. CUP 1986

McIntosh, A., 'Patterns and Ranges', in A. McIntosh & M. Halliday, *Patterns of Language*. Longman 1966

Mugdan, J., *Introduction to Morphology*. Longman, forthcoming

Murray, K. M. E., *Caught in the Web of Words*. Yale UP 1977

Nida, E. A., *Componential Analysis of Meaning*. Mouton 1975

Palmer, F. R., *Semantics* (2nd edn). CUP 1981

Potter, S., *Our Language*. Penguin 1950

Pyles, T., *The Origins and Development of the English Language* (2nd edn). Harcourt Brace Jovanovich 1971

Quirk, R., *The Linguist and the English Language*. Edward Arnold 1974

Quirk, R. & Greenbaum, S., *A University Grammar of English*. Longman 1973

Quirk, R., Greenbaum, S., Leech, G. & Svartvik, J., *A Comprehensive Grammar of the English Language*. Longman 1985

Sinclair, J. McH., 'Beginning the Study of Lexis', in C. Bazell *et al.* (eds.), *In Memory of J. R. Firth*. Longman 1966

Sledd, J. & Ebbit, W. R., *Dictionaries and THAT Dictionary*. Scott Foresman & Co. 1962

Ullmann, S., *Semantics*. Basil Blackwell 1962

Wells, R. A., *Dictionaries and the Authoritarian Tradition*. Mouton 1973

Whorf, B. L., *Language, Thought and Reality*, selected writings edited by J. B. Carroll. MIT Press 1956

Index